The Science of Monsters:
Tracking the Real-Life Creatures

by

Joe Nickell

The World's Only Professional Paranormal Investigator

MH

MONSTER HOUSE LLC

Who is Joe Nickell?

"*Nickell is the country's most accomplished investigator of the paranormal.*" —The New Yorker

"*The world's longest-running full-time professional paranormal investigator.*" —NBC News

"*Joe Nickell is a true detective of the impossible. A master sleuth that anybody who is interested in the investigation of mysteries should read and study.*" —Massimo Polidoro, Italian Psychologist

"*The dean of scientific paranormal investigation.*" —Robert Carroll

"*In a field that's commonly divided into 'believers' and 'debunkers'— people whose minds have been made up prior to inquiry—Nickell has gained international attention for being a 'fair-minded investigator.'*" —The Daily Mail

"*When it comes to investigating the paranormal, there is no one on the planet better qualified than Joe Nickell. He has the perfect skill set, having trained in the deceptive arts of conjuring and mentalism, the investigative skills of forensic science, and the communication skills of a great writer and public speaker. Over the years, he has investigated ghosts, psychics, UFOs and much more besides. In this volume, he turns his attention to monsters. Prepare to be enlightened!*" —Christopher French, British Psychologist

Acknowledgements

I am extremely grateful to John and Mary Frantz for their generous financial support for my investigative work over many years. I am especially appreciative of the efforts of Paul E. Loynes for typesetting, and to Julia Lavarnway for shepherding this manuscript through many difficulties, as well as the entire staff of *Skeptical Inquirer* magazine where much of this material appeared in an earlier form. Finally, enormous thanks to Kathleen Smith for her inestimable assistance in editing the final draft of this volume.

About the author

JOE NICKELL, PhD, has combined a background in magic (Resident Magician at the Houdini Hall of Fame) and investigation (Pinkerton "operative," forgery expert, homicide consultant) into a career solving cases by seeing through illusions and deceptions. For example, he demonstrated that the multi-humped Gloucester Sea Serpent of two centuries ago had been a group of narwhals and that a cold-case "homicide" had been a suicide staged as such so that the man's family would inherit a million dollars in insurance. Italian writer Massimo Polidoro has called Nickell "The Detective of the Impossible."

Table of Contents

Acknowledgements_____ 4

About the author_____ 5

Introduction _____ 8

Part 1: Monsters of Rivers, Lakes, and Seas _____ 9

Kraken: Monster of the Deep_____ 10

Steller's Sea Ape: Identifying an Eighteenth-Century Cryptid _____ 21

Gloucester Sea Serpent Mystery Solved after Two Centuries _____ 30

The Enigma of Cadborosaurus _____ 39

The Silly Ness Monster _____ 44

Quest for the Giant Eel _____ 52

Lake Cryptid Lookalikes_____ 62

Arkansas's White River Monster: Very Real, But What Was It? _____ 70

Gigantic Catfish: Investigating a Whopper_____ 78

Louisiana Swamp Monster _____ 88

Part 2: The Hairy Man-Beasts _____ 93

Chinese Ape-Men In Science and Myth _____ 94

Bigfoot as Big Myth _____ 105

Sasquatch Lookalikes _____ 119

Bigfoot Roundup: Some Regional Variants _____ 131

Tracking Florida's Skunk Ape _____ 143

Man-Beasts of the Yukon _____ 152

Searching for the Yowie: Australia's 'Great Hairy Man' _____ 161

Part 3: A Menagerie of Cryptids, Strange Creatures, and Supernaturals _____ **171**

The New Zealand Moa: From Extinct Bird to Cryptid _____ 172

The Giant Panda Discovered in the Land of Myth _____ 180

Solving 'Mothman' (and Others) _____ 191

Montauk Monster and the Raccoon Body Farm _____ 198

Song of a Siren: A Study in Fakelore _____ 206

The Lorelei _____ 211

Zanzibar's Popobawa: Demonic Monster That Attacks Skeptics _____ 212

Investigating Werewolves at Moosham Castle _____ 217

At a Vampire's Grave _____ 226

In Search of the 'Chupacabra' _____ 231

Identifying the Enigmatic 'Dover Demon' _____ 240

Revealing the Real Carolina Lizard Man _____ 250

Afterword _____ 257

Introduction

I have spent my life—certainly a half-century career—investigating the world's strangest mysteries: whether paranormal controversies, historical or literary enigmas, or forensic challenges.

Over the years, I have been in virtually continuous training as, first, a magician and mentalist, then a private eye with a world-famous investigative agency, and later a scholar and teacher with a doctorate in English literature and folklore.

Since 1995 I have been Senior Research Fellow with the Committee for Skeptical Inquiry, becoming apparently the world's only full-time, science-based investigator of fringe-science claims. My great friend, Italian writer and investigator Massimo Polidoro, dubbed me "The Detective of the Impossible."

In this volume, I share some of my most interesting cases involving monsters. Each of these investigations challenged my thinking and drew me further into my unique career.

Please join me in this inside look into *The Science of Monsters*. Put on your critical-thinking cap, and be prepared to sometimes think outside the box. Let us follow the evidence objectively and remember basic principles: that the burden of proof is on the claimant, not on anyone else to prove a negative; that extraordinary claims require extraordinary proof; and that the explanation that requires the fewest assumptions is most likely to be correct (the rule of Occam's razor).

Let's get started! As Sherlock Holmes (in the "Adventure of the Abbey Grange") said, waking his faithful sidekick: "Come Watson, come! The game is afoot. Not a word! Into your clothes and come!"

Part 1: Monsters of Rivers, Lakes, and Seas

Kraken: Monster of the Deep

The Kraken—a massive sea monster—legendarily rose out of the ocean to pluck sailors off ship decks or even to grasp whole vessels and carry them to the depths. It has sometimes been linked to the biblical Leviathan (e.g., Psalm 104:26; Isaiah 27:1) and the "world-serpent" of old Norse tales, Jörmungandr. It was so large, some said, it could be mistaken for an island. It was a fitting subject for Tennyson's apocalyptic sonnet, "The Kraken" (1830):

> Below the thunders of the upper deep,
> Far, far beneath in the abysmal sea,
> His ancient, dreamless, uninvaded sleep
> The Kraken sleepeth....
> (Leach 1984, 589; Levy 1999, 18–23).

It is now clear that accounts of the multi-appendaged creature best describe the giant squid, although it is only a fraction of the length attributed to the mythic beast. The giant squid's genus is *Architeuthis*, and there is only a single species. (An even larger squid—*Mesonychoteuthis* or the colossal or Antarctic squid—exists, but its range is roughly south of the tips of South America, Africa, and New Zealand [Dockett 2017; "Colossal" 2018].) Despite the identification of the Kraken as the giant squid, however, there remains much to investigate—both about the maximum size of the actual creature and the accuracy of some accounts that may be greatly exaggerated or even outright fictional.

Figure 1.1 Fanciful illustration of an enormous Kraken poised to drag a ship to the depths.

Encounters

My interest in the legendary Kraken was renewed on a trip to the Canadian province of Newfoundland and Labrador in 2008, courtesy of the *MonsterQuest* TV series to investigate Lake Crescent's purported "giant eel." (It proved to be, most likely, otters swimming in a line [Nickell 2009].) In the course of my further travels in the province, I learned of giant squid along the Newfoundland coast and sought out early accounts. I was subsequently one of the "stars" (cast) of the "Kraken" episode of Animal Planet's popular *Lost Tapes* TV series in 2010.

Historically—following specimens that were sighted or stranded at Denmark in 1545, Iceland in 1639, and Ireland in 1673—the first in North American waters was reported at Newfoundland's Grand Banks in 1785 (Ellis 1994, 125, 126, 130; Ellis 1998, 257). Other reports eventually followed.

In October 1873, two fishermen and a boy in a rowboat in Conception Bay, Newfoundland, came upon a dormant floating object. Suddenly, a terrible creature came to life and fastened one tentacle to the boat while another

11

encircled it and pulled so that water began to pour inside. Twelve-year-old Tom Piccott had the presence of mind to grab up a "tomahawk" from the bottom of the boat and chop at the estimated thirty-five foot tentacle. When he had hacked it off, he did the same with the other attached to the little boat, and the creature retreated into the ocean while emitting an inky fluid to obscure its trail.

Tom took the longer piece of tentacle to St. John's to a Presbyterian minister and amateur naturalist named Rev. Dr. Moses Harvey who had long been fascinated by tales of the "Devil Fish." He paid Tom ten dollars for the monstrous tentacle, later recalling:

> How my heart pounded as I drew out of the tub in which he carried it, coil after coil, to the length of nineteen feet, the dusty red member, strong and tough as leather, about as thick as a man's wrist. I knew at a glance it was one of the tentacles or long arms of the ancients' Kraken, or modern giant cuttle fish. Eureka! (Harvey 1899).

Harvey's interest and expertise led him to further luck. Just three weeks later he was called to Logy Bay where an almost complete specimen had been brought, entangled in some fishermen's net. As they tried to drag the net to shore, tentacles extending through it tried to grasp the boat, whereupon one man drew his splitting knife and severed most of the creature's head. Of course, Harvey bought this specimen too, took it home in his wagon, and stored it in brine in his shed. Although this creature was smaller than the previous one, its tentacles measured up to twenty-four feet.

Harvey presented his giant squid specimens and papers to Yale University professor A.E. Verril, an authority on cephalopods. Four years later, a severe windstorm drove another giant squid ashore at Catalina, and it was acquired by the New York Aquarium for $500. Still other encounters and specimens proliferated (Fitzgerald 2006, 49–71).

Attack on a Schooner

As striking as these encounters are, it was the giant squid's attack on a 150-ton schooner in 1874 that reportedly spilled—not squid "ink," but copious printing ink in newspapers throughout Canada and the United States. My source cited testimony of both many survivors and witnesses from a nearby ship, the *Strathowen*. Supposedly the schooner, the *Peril*, had been in Atlantic waters to the south of Newfoundland.

Suddenly a gigantic creature rose from the sea and wrapped its tentacles around the *Peril*. Captain James Flood later told how he had grabbed up his rifle and—despite Newfoundlander Bill Darling's warning not to shoot and so enrage the huge creature—did just that. "The oblong body," he said, "was at least half the size of our vessel in length and just as thick. The train must have been a hundred feet long." Although the men assailed the monstrosity with axes, it pulled the craft over and dragged it beneath the waves. The *Strathowen* moved in then and rescued the captain and his surviving crew members.

To learn more, I followed my source (Fitzgerald 2005, 74–76) to *his*, an account by Edward Rowe Snow in Snow's *Mysteries and Adventures Along the Atlantic Coast* (1948). To my consternation, *Peril* is there named as "*Pearl*" and Capt. Flood as "Floyd," the location of the encounter is not given, and the creature is described not as a squid but as "a giant octopus"! Moreover, the original news source is not Fitzgerald's *News World* but instead "*News of the World* for July 5, 1874."

Seeking to verify the story and to document the facts, I set Google on the trail and in time was led from waters south of Newfoundland across the Atlantic and around the tip of Africa to the Indian Ocean's Bay of Bengal! From there, in June 1874 reportedly, accounts supposedly appeared in Indian newspapers about an attack on a 150-ton schooner—accounts whose text was nearly identical to the foregoing.

In fact, the publishing trail soon led to *The Times* of London, where intrepid CFI Libraries Director Tim Binga was eventually able to secure a digital copy of the story, titled "A Successor to the Sea Serpent," which appeared in

The Times of July 4, 1874, on page 8. However, it had in turn been taken from the *Homeward Mail*, a British paper in India, and we soon had that text of Monday, June 29, 1874, in hand.

The *Homeward Mail's* account is appropriately headed, "A Very Strange Story," and purports to tell how some unidentified person aboard the steamer *Strathowen*, having witnessed the attack on the schooner, "prevailed on the skipper," after his rescue, "to give me his written account." That follows in quotation marks and is signed, "James Floyd, late master schooner *Pearl*."

Truth Be Told

But is this story true? I immediately found it very doubtful, but some others have accepted it. The authority on the giant squid, Richard Ellis (1998, 257–265), includes it in his list of 121 "Authenticated Giant Squid Sightings" (1545–1996); however, his source is at a century's remove. Fitzgerald (2005, 74–76) insists of the attack on the schooner, "This attack was not imaginary"; and yet we have seen that he thought it occurred off the Newfoundland coast, and his source was also untrustworthy (Ellis 1998).

No less a figure than Arthur C. Clarke wrote of *The Times* account, "If you think this an improbable story, I do not blame you; I would not take it very seriously myself, if it had appeared anywhere except in the shipping column of *The Times*—a paper not noted for sensational journalism" (Clarke 1974). However, as we have seen, it did first appear somewhere else, in a source lacking *The Times'* reputation, *Homeward Mail*, which said it had in turn been "communicated to the Indian papers." By whom is not stated, so the original remains unknown—three or so steps back from *The Times*. I feel sure that if Clarke had known all this he would not have given his endorsement— equivocal though it was.

The evidence regarding the credibility of the tale is bad indeed. There is no record of a ship named the *Strathowen* in Lloyd's Register or other source, and a review of the account more recently (Boyle 1999) laments, "It is unfortunate that the facts concerning the encounter in question have never

14

been wholly substantiated."

Moreover, the Bay of Bengal, the northern extension of the Indian Ocean, is not a known habitat of giant squid, based on the locations of beaches they have washed up on. They are rarely found in the tropics (Smithsonian Ocean 2018).

From what we know of the giant squid, such an attack as that alleged on a 150-ton schooner would not seem credible. The largest specimen known was only fifty-nine feet long and weighed just under a ton—being a female, which is typically larger than the male ("Giant Squid Facts" 2018). Regarding old sailors' tales of giant squids overpowering ships or snatching someone from deck or shore, advises one source, "None of these turn out to be true" ("Giant Squid Found" 2007).

That the writer is not only anonymous but avoids telling us whether he is a reporter or other communicant is suspicious, as is the introductory phrase that the story "has been communicated"—the passive-voice construction aiding in the concealment of the author. And of course the existence of "Captain James Floyd" likewise remains unverified and questionable.

"Floyd" certainly seems to have missed his true calling—not as a ship's captain but as a spinner of adventure yarns. He reproduces dialect, having one crewman state, "and it ain't the sea sarpent, for he's too round for that ere critter." His attention to detail is admirable: he notes that three of the crew "had found axes, and one a rusty cutlass." He describes "the advancing monster" as "a huge oblong mass moving by jerks just under the surface of the water." Then, he says, "In the time I have taken to write this the brute struck us, and the ship quivered under the thud; in another moment, monstrous arms like trees seized the vessel and she heeled [tilted] over." I suspect "Floyd" is the invention of the anonymous author.

The concocted narrative may have been inspired by a similar episode in Jules Verne's novel *Twenty Thousand Leagues Under the Sea* published in 1869. In it a giant poulp (a cephalopod such as an octopus or squid) has become entangled in the *Nautilus*'s screw, and Captain Nemo raises the submarine to the surface. Opening the hatch, he and the crew use axes to battle, not just one creature but, over a quarter hour's time, ten or twelve from a "nest of serpents, that wriggled on the platform in the waves of blood and ink." (One of the crew is Ned Land, a Canadian, who may have been the model for the *Pearl*'s Bill Darling, a Newfoundlander.)

Conclusions

Beyond mythology and fantasy, the Kraken of old is of renewed interest as the giant squid (*Architeuthis*)—diminished in size but authentically real. Even at "only" sixty feet, it rivals the sperm whale for length, and we do not really know how large it may get. Richard Ellis, in his definitive book (1998, 8), states: "It is the least-known large animal on earth, the last monster to be conquered."

Notes

1. Squid have *eight arms*, which have suckers along their entire length, and a *pair of longer tentacles*, with suckers only at the tip.

2. A list of squid sightings and strandings (Ellis 1998, 257–264) lists only two for the Indian Ocean: one the questionable *Pearl* story and one other, the following year, in the Southern Indian Ocean, of "unknown" size.

References

Boyle, Richard. 1999. When the squid sank a schooner. *The Sunday Times* (December 12).

Clarke, Arthur C. 1974. *The Treasure of the Great Reef*; quoted in Boyle 1999.

Colossal Squid. 2018. Available online at https://en.wikipedia.org/wiki/Colossal_squid; accessed September 6, 2018.

Dockett, Eric. 2017. Colossal Squid vs. Giant Squid: The Real Kraken Sea Monster. Available online at https://owlcation.com/ stem/Colossal-Squid-vs-Giant-Squid-the-Real-Kraken-Sea-Monster; accessed September 6, 2018.

Ellis, Richard. 1994. *Monsters of the Sea*. New York: Alfred A. Knopf.

———. 1998. The *Search for the Giant Squid: The Biology and Mythology of the World's Most Elusive Creature*. New York: Penguin Books.

Fitzgerald, Jack. 2005. *Ghosts and Oddities*. St. John's, NL Canada: Jefferson Publishing.

———. 2006. *Newfoundland Adventures*. St. John's, NL Canada: Creative Publishers.

Giant Squid Facts. 2018. Available online at www.softschools.com/facts/animals/giant_squid_facts/113; accessed September 18, 2018.

Giant Squid Found. 2007. *MonsterQuest*, Season 1, Episode 3, originally aired November 14.

Harvey, Rev. Moses. 1899. A Sea Monster Unmasked. *Science Digest*. Cited in Fitzgerald 2006.

Leach, Maria, ed. 1984. *Funk & Wagnalls Standard Dictionary of Folklore, Mythology, and Legend*. New York: Harper & Row.

Levy, Joel. 1999. *A Natural History of the Unnatural World*. New York: St. Martin's Press.

Nickell, Joe. 2009. Quest for the giant eel. *Skeptical Inquirer* 33(4) (July/August).

Smithsonian Ocean: Giant Squid. 2018. Available online at https://ocean.si.edu/ ocean-life-/invertebrates/giant-squid; accessed September 18, 2018.

Snow, Edward Rowe. 1948. *Mysteries and Adventures Along the Atlantic Coast*. Edition updated by Jeremy d'Entremont; Carlile MA: Commonwealth Editions, 2004, 133–134.

A Successor to the Sea Serpent. 1874. *The Times* (London) (July 4): 8.

Verne, Jules. 1869. *Twenty Thousand Leagues Under the Sea*. English translation, Project Gutenberg ebook, updated 2016, chapter XVIII: "The Poulps" (pp. 127–147).

A Very Strange Story. 1874. *Homeward Mail from India, China and the East* (June 29)

Steller's Sea Ape: Identifying an Eighteenth-Century Cryptid

Since its appearance in 1741, a mysterious creature has remained controversial—a so-called "sea monkey" that puzzled naturalist Georg Wilhelm Steller. He encountered the mystery creature during his service as physician and scientist on Commander Vitus Bering's Second Kamchatka Expedition to America, 1741–1742.

Some cryptozoologists (active in a field that postdates Steller) refer to the creature as a marine "cryptid" (an animal whose existence—or current existence—is scientifically unverified and thus of interest to cryptozoologists). They use the name Steller's Sea Ape ("Georg Wilhelm Steller" 2013; Coleman and Huyghe 1999, 64–65; Mackal 1983, 1–32). I determined to approach the case as a skeptical cryptozoologist—more specifically as what I term a *paranatural naturalist* (one who first considers allegedly paranatural entities as hypothetically natural creatures and then seeks to identify them [Nickell 2015]). I determined to conduct a new investigation.

Steller's Description

Steller (1709–1746) wrote an account of his sighting in his journal at the time. After his early death this passed to the archives of the Russian Academy of Sciences (a body I was once honored to speak before). There it was discovered nearly a century later and published (Steller 1793). I quote from the second (and best) of two English translations of the German text (Steller in Golder 1925, 64–66).

Steller notes that they were at about 52.5° N latitude, 155° W longitude—off the Shumagin Islands of the Aleutians (that constitute southwest Alaska). As he wrote:

On August 10 we saw a very unusual and unknown sea animal, of which I am going to give a brief account since I observed it for two whole hours.—It was about two Russian ells in length [i.e., about five feet long]; the head was like a dog's, with pointed, erect ears. From the upper and lower lips on both sides whiskers hung down. The eyes were large; the body was long, rather thick and round, tapering gradually towards the tail. The skin seemed thickly covered with hair, of a gray color on the back, but reddish white on the belly; in the water, however, the whole animal appeared red, like a cow. The tail was divided into two fins, of which the upper, as in the case of roosters, was twice as large as the lower. Nothing struck me as more surprising than the fact that neither forefeet nor, in their stead, fins were to be seen.

He likened the creature to one that another naturalist, Konrad von Gesner, had reported, and so called it "Gesner's sea monkey." He continued:

For over two hours it swam around our ship, looking, as with admiration, first at the one and then at the other of us. At times it came so near to the ship that it could have been touched with a pole, but as soon as anybody stirred it moved away a little farther. It could raise itself one-third of its length out of the water exactly like a man, and sometimes it remained in this position for several minutes. After it had observed us for about half an hour, it shot like an arrow under our vessel and came up again on the other side; shortly after, it dived again and reappeared in the old place; and in this way it dived perhaps thirty times. There drifted by a seaweed, club-shaped and hollow at one end like a bottle and gradually tapering at the other, towards which, as soon as it was sighted, the animal darted, seized it in its mouth, and swam with it to the ship, making such motions and monkey tricks that nothing more laughable can be imagined. After many funny jumps and motions it finally darted off to sea and did not appear again. It was seen later, however, several times at different places of the sea.

The first thing to emphasize is that "there is no known animal, in the sea or, for that matter, on land, that corresponds to Steller's description,"

22

according to the late cryptozoologist Roy Mackal (1983, 5). Some have thought Steller had seen some known sea creature such as a sea otter or fur seal (Mackal 1983, 6). Yet as a naturalist of note who identified several species of North American plants and animals, Steller was a skilled observer who had an excellent opportunity to see the animal for two hours, although its several close-up appearances were seemingly each brief.

It seems unlikely to have been a hoax, although that has been suggested, or a satire: Since Steller's text uses the term *Simia marina danica*, "Danish Sea Ape," the thought is that Steller was poking fun at Captain Bering, "the only Dane on the ship" (Thaler 2011). However, the description contains no hint of satire and would seem more likely to have been an honorific naming.

Toward a Solution

We are looking, therefore, for a *real creature*—one that, as Mackal notes (1983, 19, 26), was "most certainly" a mammal, given its anatomy and movements, and "most probably" a *seal*, which it most resembled. So far, so good, but would Steller not know a seal when he saw one?

There are two families of seals (which, with the Walruses, constitute the order Pinnipedia). The eared seals (family Otariidae)—consisting of the smooth-coated sea lions (the "trained seals" of circus acts) and the fur seals—have small though noticeable external ears and longer necks than seals of the other family. Earless seals (family Phocidae), also called hair seals or true seals, have only a small orifice in place of an external ear (Whitaker 1996, 712–744).

In discussing Steller's Sea Ape, Mackal correctly singles out the element of the apparently missing forelimbs or fins as crucial. Such a unique instance of congenital deformity in a seal would have been extremely statistically unlikely. More plausible was that when the animal swam it held its fore flippers pressed closely against its body—as in fact the Leopard Seal does. Unfortunately, the Leopard Seal lives in the southern hemisphere, and it has no external ears. Nevertheless, Mackal (1983, 30) suggested the possibility of an unknown seal species having evolved with characteristics of Steller's creature: having both ears and a tendency to swim in the fashion described (probably with flippers folded rather than missing). Mackal thought a new expedition might well discover just such an aquatic mammal.

But Mackal, I believe, should have considered that his hypothesis seems even more statistically unlikely than that of a seal with missing fore flippers. Moreover, his approach assumes that, if we cannot identify the creature, it must be a hitherto unknown one.

Identification

I took a different tack, considering it more likely that Steller's observation was imperfect. He admitted as much when he explained his reasons for firing on the creature: it was "in order to get possession of it for a more accurate description."

Of course, what I think is the main error in Steller's description involves portions of the creature's anatomy that would have been under water or, when not, would have been viewed at a distance. First were the "missing" fore flippers. Actually, he did not insist that they were absent but instead that they were not "to be seen"—not precisely the same thing. The other feature was what Steller mistakenly thought was a tail fin. More on both of these features presently.

Allowing for these misperceptions by Steller, the animal he remotely studied was most like a Northern Fur Seal (a.k.a. "Alaska Fur Seal"). Its scientific name is *Callorhinus ursinus*, which means "the bearlike animal with the beautiful hide" (Bonner 1999, 50). In addition to having noticeable ears, it does have a doglike head, "large eyes," and "long whiskers" (Whitaker 1996, 713), hanging down, just as Steller described. It has the requisite body shape and is covered with thick hair—again like Steller's mystery creature. The long-necked fur seal could also raise a third of its length or so out of the water. As to its antics, I believe that, as described, these were within the range of seals generally and the fur seal in particular (Whitaker 1996, 713–715; Stejneger

1936, 280, 281). As it happens, the pinnipeds are characterized by their "playful disposition" and the fact, "In the wild state, they show little fear of man" (*Larousse* 1975, 546)—descriptions that tally with Steller's mystery sea animal.

Now, I made this tentative identification after reading Coleman and Huyghe (1999, 64–65), using mostly the *National Audubon Society Field Guide to Mammals: North America* (Whitaker 1996). Subsequently, Mackal's book (1983) helped convince me I was on the right track, but when I finally accessed a copy of Leonard Stejneger's 1936 biography of Steller, I realized that Stejneger had effectively solved the mystery himself. I had done little more than independently confirm his prior identification. However, since cryptozoologists had ignored or dismissed Stejneger—whose evidence was not as fully developed as it could be—I determined to press on.

More Precisely . . .

Why would Steller not have recognized a fur seal? As his biographer insists (Stejneger 1936, 281), Steller, "at the time he made his observation, *had never seen a fur-seal,* dead *or alive*" (original emphasis). When he did later make the animal's acquaintance at Bering Island, the die was already cast.

I question Stejneger on one significant point. He believes Steller saw "a full-grown bachelor fur seal." However, in two respects—size and coloring—adult males tend to be unlike Steller's mystery creature. Steller's biographer Dean Littlepage (2006) has therefore tweaked the identification to suggest a young member of the species.

However, the females, as we shall see, are an exceptionally good match. They had previously been wintering in California (along with pups and young males). Still today, they head north in spring to the Aleutians where the bulls collect large harems on rocky islands. The three months of family life (including giving birth to pups conceived the summer before, then mating again) end in early August when the females begin making "feeding forays" (Whitaker 1996, 714–715; Jordan 1952, 149).

Moreover, the female has different coloring than the male, who is blackish on its upper side, grayish on its massive shoulders, and reddish beneath. In contrast, the female has coloring like Steller's creature, which had "a gray color on the back, but reddish white on the belly." The female Northern Fur Seal similarly has "gray above, reddish below" (Whitaker 1996, 715).

Size is also important, since the Northern Fur Seal exhibits "pronounced sexual dimorphism in size," beginning at birth (Reeves et al. 1992, 50). The male is typically much larger than Steller's creature, measuring in length 6'3" to 7'3"(1.9–2.2m). On the other hand, the female is in the range of 3'7" to 4'7" (1.1–1.4m), much more in keeping with Steller's estimate of about five feet. Of course size and coloring of individuals vary, but, taken together, I think the evidence is stronger for a female.

In any event, either female or male would have shared the other features of Steller's creature: the doglike head, the drooping whiskers, large eyes, etc. If it be quibbled that the fur seal's ears are, while pointed, not really "erect" (they are actually directed backward), Steller did in fact describe those of the actual fur seal using the very same descriptors: In his *De Bestiis Marinis* ("Of Sea Beasts") he says the fur seal's ears are "*acutae . . . et erectal*" (Steller 1751, 2: 334).

And to return to the issue of the "missing" forelimbs, Steller's own failure to observe them on his creature actually argues in favor of its having been a Northern Fur Seal. That is because the position of the forelimbs on that animal is *farther back*, emphasizes Stejneger (1936, 280), "than in any related animal with which he was then familiar." Surely, that is why the creature, able to raise a third of itself upright out of the water for a few minutes (obviously treading water), failed to show forelimbs. "Moreover," adds Stejneger, "when moving at high speed through the water the fur-seal keeps the fore flippers pressed very close to the body so that they are practically invisible."

The identification of the creature as the Northern Fur Seal also helps explain Steller's other serious error, the description of the creature's tail as "divided into two fins." In reality, this was simply a misperception of the seal's

26

two closely set rear legs with their flippers (Stejneger 1936, 280).

Conclusions

In rounding up the usual suspects, we had to be careful that our net not miss what we can now see is the closest match to Steller's Sea Ape: the female Northern Fur Seal. In contrast to the male, she better fits the profile of Steller's mystery creature. She is, I believe, not only the nearest match of any existing creature but represents the preferred hypothesis in the case, according to the principle of Occam's razor (that the hypothesis with the fewer, smaller assumptions is to be preferred).

I have accepted Steller's description on a dozen points, only questioning his observation regarding a few details. I think the timing of his sighting is corroborative, occurring when the female—having the size and coloration he described—would have come from the breeding grounds.

I am glad Steller's two shots fired at the delightful animal missed. Had they not, we would have lacked the intriguing mystery she gave us. I think I'll name her Agatha.

References

Bonner, Nigel. 1999. *Seals and Sea Lions of the World*. London: Blandford.

Coleman, Loren, and Patrick Huyghe. 1999. *The Field Guide to Bigfoot, Yeti, and Other Mystery Primates Worldwide*. New York: Avon Books.

Georg Wilhelm Steller. 2013. Available online at http://en.wikipedia.org/wiki/Georg_Wilhelm_Steller; accessed December 23, 2013.

Golder, F.A., ed. 1925. *Bering's Voyages: Account of the Efforts of the Russians to Determine the Relation of Asia and America*. Reprinted New York: Octagon Books, 1968.

Jordan, E.L. 1952. *Hammond's Nature Atlas of America*. New York: C.S. Hammond and Co.

Larousse Encyclopedia of the Animal World. 1975. New York: Larousse & Co.

Littlepage, Dean. 2006. *Steller's Island: Adventures of a Pioneer Naturalist in Alaska*. Seattle: Mountaineers Books. Cited in "Steller's" 2016.

Mackal, Roy P. 1983. *Searching for Hidden Animals*. London: Cadogan Books.

Nickell, Joe. 2015. Monster lookalikes: Reflections of a paranatural naturalist. *Skeptical Inquirer* 39(3) (May/June): 12–15.

Reeves, Randall R., Brent S. Stewart, and Stephen Leatherwood. 1992. *The Sierra Club Handbook of Seals and* Sirenians. San Francisco, CA: Sierra Book Club.

Stejneger, Leonard. 1936. *Georg Wilhelm Steller: The Pioneer of Alaskan Natural History*. Cambridge, Mass: Harvard University Press.

Steller, George Wilhelm. 1751. *De Bestiis Marinis*, in 2 vols. Cited in Stejneger 1936, 280 (n.33: 285–286).

———. 1793. *Reise von Kamtscatak nach Amerika mit dem Cammandeur-Captain Bering . . .*; reprinted in English translation in Golder 1925, 9–187.

Steller's Sea Ape. 2016. Available online at https://en.wikipedia.org/wiki/Steller%27s_sea_ape; accessed March 23, 2016.

Thaler, Andrew David. 2011. Unraveling the mysteries of Steller's Sea Ape. Available online at www.southernfriedscience.com/unraveling-the-mysteries-of-stellers-sea-ape/; accessed March 23, 2016.

Whitaker, John O., Jr. 1996. *National Audubon Society Field Guide to North American Mammals*, rev. ed. New York: Alfred A. Knopf.

Gloucester Sea Serpent Mystery Solved after Two Centuries

A "wonderful sea-snake" was repeatedly seen in the area of Gloucester Bay and Nahant Bay, Massachusetts, in August 1817 and again in 1819. Although attracting "hundreds of curious spectators," plus a large reward for "his snakeship" alive or dead, the great creature escaped any such fate (Drake 1883, 156–159). The visitations have been reported in many respectable publications—including Richard Ellis's *Monsters of the Sea* (1994, 48–55, 362)—and have prompted this assessment: "Whatever this animal may or may not have been, the fact remains that it is one of the most scientifically respected encounters in the annals of cryptozoology and remains one of the great unsolved mysteries of the sea" (Morphy 2010).

Is it possible now, after two centuries, that we might actually solve the enigma? What might such a solution look like—as it begins to come dimly into view? Will it disappoint or fascinate? Or will it simply represent legend, superstition, and eyewitness error, corrected by a detective approach and access to modern research methods?

Sea Serpent!

The late, great cryptozoologist Bernard Heuvelmans (1968, 149) captures something of the developing excitement over the Gloucester monster's sudden appearance:

On 6 August 1817 two women saw a sea-monster like a huge serpent come into the harbor of Cape Ann which lies north of Gloucester roads. Little attention was paid to their story, although it was confirmed by several fishermen, but a week later so many people known to be

30

trustworthy claimed to have seen the animal nearby, that the whole country round was much excited. On 10 August a seaman called Amos Story saw it from the shore. It was near Ten Pound Island in the shelter of Gloucester roads. On 12 August, Solomon Allen 3rd, a shipmaster, saw it from a boat, and again during most of the following day, and for a short time on the fourteenth, when it was watched by twenty or thirty people, including the Justice of Peace of Gloucester, the Hon. Lonson Nash. On that day four armed boats were sent in pursuit of the monster, and Matthew Gaffney, a ship's carpenter, fired at it at almost point-blank range, apparently hitting it with a musket ball in the head, but doing it no harm.

Heuvelmans concludes, "There was no doubt about it, the fabulous sea-serpent was there, large as life, in Gloucester harbour."

A broadside that appeared on August 22—titled "A Monstrous Sea Serpent: The largest ever seen in America"—provided important information, illustrating how observations were interpreted in light of the prevailing belief that such marine creatures were serpentine.

It read (Ellis 1994, 49–50):

There was seen on Monday and Tuesday morning playing around the harbor between Eastern Point and Ten Pound Island, a SNAKE with his head and body about eight feet out of the water, his head is in perfect shape as large as the head of a horse, his body is judged to be about FORTY-FIVE to FIFTY FEET IN LENGTH. It is thought that he will girt about 3 feet round the body, and his sting is about 4 feet in length.

It was first seen by some fishermen, 10 or 12 days ago, but it was then generally believed to be a creature of the imagination. But he has since come within the harbor of Gloucester, and has been seen by hundreds of people. He is described by some persons who approached within 10 or 15 yards of him, to be 60 or 70 feet in length, round, and

31

of the diameter of a barrel. Others state his length variously, from 50 to 100 feet.

What Was It?

That the sea serpent went unidentified was not for lack of effort. Naturalist Constantin Rafinesque gave it his attention and believed with others that it was "evidently a real sea snake" (Ellis 1994, 54). The New England Linnaean Society even tried to palm off a new scientific designation: *Scoliophis Atlanticus* (Morphy 2010). A Captain Rich threw a harpoon into the monster as it swam under his whaleboat, but the iron pulled out after fifty yards. Among modern investigators, Richard Ellis (1994, 362), upon studying the case at length, asked with some exasperation, "Shall we assume that hundreds of reputable citizens were deluded or victims of mass hysteria?" Still later, a lady writing a novel had a creative idea: maybe the creature was a "poor humpback whale, entangled with a net or rope lined with keg or cork buoys" (Fama 2012).

If we hope to do better, we should begin by questioning the dogma that the creature was a great snake. In fact, as the Boston *Centinel* took pains to note, it did not "wind laterally along, as serpents commonly do, but his motion is undulatory, or consisting of alternating rising and depression." Other sources, including an affidavit by a ship's carpenter, agreed that "his motion was vertical" (Ellis 1994, 50, 51). In other words, the movement was not sideways like that of a reptile but undulating like a cetacean (a marine mammal), and that should be one of our first clues as to the creature's identity.

Furthermore, on the sixteenth, four men at first saw the monster—not as a single serpentine creature but instead as a school of pilot whales. But, likely influenced by the prevailing view, they then came to think these "whales" looked collectively like the humps of a sea monster (Ellis 1994, 50).

This is a well-known illusion, as, for instance, multiple otters can be mistaken for a single, long, multi-humped creature (Nickell 2007).

The tendency to "see" a serpentine form, by connecting the dots (or humps), seems powerful. Several eyewitnesses to the Gloucester monster got different counts: e.g., thirteen, fifteen, twenty, and thirty-two humps, thus getting overall length estimates of 100 or 120 feet or more (Ellis 1994, 50–54; Heuvelmans 1968, 165).

But what about another "serpentine" feature mentioned briefly in the Boston broadside of 1817? Recall the statement that the great snake's "sting" (or stinger) was "about 4 feet in length." The term *sting* was once used historically to mean the wound from the *stinger* of a snake, as in "the sting of an asp." (Now snakes are correctly said to bite, not sting.) Indeed, in one witness's words, the serpent "threw his tongue [i.e., his stinger] backwards several times over his head" (qtd. in Soini 2010, 107).

This archaic reference is telling, because there is indeed a marine mammal with a long tusk-like member (in fact a canine tooth) that extends from the head to a length of about 4.9 to 10.2 feet during the up-to-fifty-year life of a male. (A small percentage of females of the species also grow a tusk, but it is smaller.) It appears that this was the four-foot "sting" or "stinger" that emanated from the Gloucester sea serpent's head, mistaken for a tongue, and thus helps in identifying it at last as the narwhal (or narwhale), which uniquely has this feature.

A New Identification

The narwhal is a medium-sized whale—one of two living species in the family *Monodontidae* (the other being the beluga whale). They range in size from about thirteen to eighteen feet. They tend to congregate in groups of five to ten or even twenty or more. Groups may consist of females and young (about six feet long) or also contain juveniles and adult males. At Gloucester, we can suspect that such a group was mistaken once for a school of pilot whales, then for a single great, multi-humped sea serpent, with eventually one male exhibiting a perceived "stinger." (That horny feature extending from the head has earned the narwhal the nickname "unicorn of the ocean," which is all the more apt because it is "amongst the world's rarest whales" ["Animalia" 2019].)

Indeed, I sought more evidence of the reported stinger—aware that, despite its relatively small diameter, which could make it difficult to see, more than one individual must have possessed one. In time I came upon another—mentioned in affidavits collected by the Linnaean Society (Report 1817, 153). One witness had reported seeing an open mouth, while another described, extending from the front of the creature's head, a long pointed "prong or spear" extending two feet or so from its jaws (Report 1817, 153; Soini 2010, 107). Also, yet another witness had seen such a feature, about twelve inches long (Soini 2010, 91). I take these instances as providing clear, corroborative evidence for the existence of the narwhal's horn-like tusk—observed on different individuals. Because the tusk makes narwhals unique among marine animals, it represents extremely strong evidence that the "creature" was actually a group of narwhals—however confusingly perceived.

Another element corroborating the identification of narwhals is found in the fact that in August 1819, the Gloucester Sea Serpent appearing at Nahant, Massachusetts, "was stationary for four hours near the shore (qtd. in Ellis 1994, 54). This is quite consistent with narwhal behavior. Its name derives from the Old Norse *nár* (*corpse*) that refers both to its greyish, mottled coloration as well as its summertime habit of lying still (called "logging") on the surface of the water—whether in this case as an individual or "pod" ("Narwhal" 2019; "Animalia" 2019).

Among other features consistent with a narwhal was the creature's size—an important point because witnesses were seeing only above-water portions and failed to realize whether they had seen one or multiple individuals. Also, the coloration was typically described as "dark" above and "nearly white" on the underbelly (as far as could be seen); old narwhals eventually become almost totally white ("Animalia" 2019).

34

A particularly distinctive feature was the creature's ability, described by several witnesses, "to dive straight down without twisting its body in the least" (Report 1817, 153). Heuvelmans (1968, 153) observed that this could be explained if the Gloucester monster had fins or flippers on its sides (hidden under the water) as, of course, the narwhal does, having short, rounded flippers. Also, like narwhals, the monster could swim at great speed (Report 1817, 149–155). Narwhals are "unique and amazing swimmers" and often swim belly up, being capable of sudden, fast, deep dives ("Animalia" 2019).

As we have seen, descriptions of the Gloucester serpent varied widely—not surprisingly because some saw one relatively closely, while others viewed "it" from afar. One witness, for example, watched through his spyglass at a quarter of a mile's distance. He, like most others, saw no eyes, mane, breathing holes, fins, or other fine features. Like many others, however, he reported the creature did have its head out of the water—as the narwhal often does. Most who provided testimony insisted the animal seemed quite unaggressive (Heuvelmans 1968, 152–153), again like narwhals.

Conclusions

The narwhal scenario I have postulated here would seem to explain why such a perceived sea serpent appeared—but only did so rarely—in Gloucester Bay, leaving a great mystery in its wake. The prevailing conviction that sea serpents were great multi-humped creatures led eyewitnesses to mistake a "pod" of smaller narwhals—which the witnesses were unfamiliar with—for a monstrous serpent.

Most likely, however, is the explanation from an earlier, unique shift in ocean temperature that I missed (but that was noted by some astute readers of my original article in *Skeptical Inquirer* magazine—the first of whom was Tom Flynn). The year before the first incident (i.e., 1816) came a period well known to oceanographers and weather experts as "the year without a summer" (see "Year without a Summer" 2021).

This was an occurrence of notably cold weather (coincidentally at the end of the little ice age), due to the tremendous volcanic eruption of Indonesia's Mount Tambora in 1815, which blocked out sunlight. In effect, it caused the Arctic Circle to extend south to Gloucester Bay, thus extending the narwhals' range—for just the time period in question.

Gloucester would have been south of narwhals' expected range today, which includes the Atlantic area of the Arctic Ocean. Individuals are often found in the Canadian Arctic Archipelago—the northern part of Hudson Bay, for example. Narwhals do exhibit seasonal migrations, moving closer to coasts in summer months ("Narwhal" 2019). "Narwhals usually do not stray far below the Arctic Circle," but stragglers have been recorded around Newfoundland, Europe, and the eastern Mediterranean (Drury 2019). It would not seem far-fetched for a group to have appeared as far south as Gloucester in 1817 and 1819. I think the distinguishing feature of the monster's "sting," "prong," or "spear"—unique to the narwhal—argues that is just what happened.

Today, like many mammals, narwhals are threatened by human activity. They are among those Arctic marine animals that are most vulnerable to climate change, due to reduced sea ice coverage in their environment. What an irony that two centuries after humans gathered to watch the mysterious Gloucester Sea-Serpent (and yes, attempted to kill it) we have learned its true nature—without perhaps quite yet learning our own.

References

Animalia: Narwhal. 2019. Available online at
    http://animalia.bio/narwhal; accessed May 6, 2019.

Drake, Samuel Adams. 1883. *New England Legends and Folklore*. Reprinted New York: Chartwell Books, 2017.

Drury, Chad. 2019. Monodon monoceros narwhal. Available online at
    https://animaldiversity.org/accounts/Monodon_monocer os; accessed May 9, 2019.

Ellis, Richard. 1994. *Monsters of the Sea*. New York: Alfred A. Knopf.

Fama, Elizabeth. 2012. Debunking a Great New England Sea Serpent. Available online at
    https://www.tor.com/2012/0816/debunking-a-great-new-england-sea-serpent; accessed May 7, 2019.

Heuvelmans, Bernard. 1968. *In the Wake of the Sea-Serpents*. Translated from the French by Richard Garnett. New York: Hill and Wang.

Morphy, Rob. 2010. Gloucester Sea Serpent (Massachusetts, USA). Available online at
    https://www.cryptopia.us/site/2010/02/gloucester-sea-serpent-massachusetts-usa; accessed April 30, 2019.

Narwhal. 2019. Available online at
    https://en.wikipedia.org/wiki/Narwhal; accessed May 2, 2019.

Nickell, Joe. 2007. Lake monster lookalikes. *Skeptical Briefs* 17(2) (June): 6–7.

O'Neill, J.P. 2003. *The Great New England Sea Serpent*. New York: Paraview Special Editions.

Report of a Committee of the Linnaean Society of New England. 1817. Portions cited and quoted from affidavits taken from several eyewitnesses by The Hon. Lonson Nash et al., in Heuvelmans 1968, 149–155. (See also O'Neill 2003, 43–44, 51.)

Soini, Wayne. 2010. *Gloucester's Sea Serpent.* Charleston, SC: History Press.

Year without a Summer. 2021. Available online at https://en.wikipedia.org/wiki/Year_Without_a_Summer; accessed September 15, 2021.

The Enigma of Cadborosaurus

Mankind's imagination has always been excited by the possibilities of unknown regions. Thus, a seemingly limitless universe invites speculation about extraterrestrials; the world's largely unexplored oceans and seas, even deep lakes, prompt thoughts of leviathans; similarly, vast wilderness areas of the globe spark belief in other strange creatures.

In mid-2006, I was aboard a Center for Inquiry cruise that traveled north from Seattle, Washington, along the coastal reaches of British Columbia and southern Alaska. As part of our floating conference on "Planetary Ethics" I spoke on "Mysterious Entities of the Pacific Northwest," which I specially researched for the cruise, and—as opportunity presented itself—I was also able to do a bit of on-site investigating relating to that topic as we occasionally put into port. It was not wasted effort, as we shall see.

Sea Serpents?

That there are—if not actual "sea serpents"—great denizens of the deep, no one can dispute. Among them are the giant manta ray (frequently twenty feet across), the whale shark (sixty or more feet long), and still other great creatures—including the giant squid and the blue whale (Welfare and Fairley 1980, 68, 71–72).

While there are numerous early accounts of great "sea serpents," often described as having multiple humps, it is usually difficult to theorize about what was actually seen. In one instance it may have been quite ordinary creatures viewed at a distance, or in another simply the product of an overworked imagination or even a deliberate tall tale. The lack of photographs is one problem, the absence of a single authenticated remnant another.

There are *apparently* such remains, such as the carcass of one that washed ashore in Scotland in 1808 (known as the Stronsa Beast) and another caught in a Japanese fishing net on April 25, 1977 (Welfare and Fairley 1980,

81; Shuker 1996, 210–211). Both of those turned out to be the rotting carcasses of basking sharks. According to *Arthur C. Clarke's Mysterious World*: "The dead basking shark decays in the most deceiving manner. First the jaws, which are attached by only a small piece of flesh, drop off leaving what looks like a small skull and thin serpentlike neck. Then as only the upper half of the tail fin carries the spine, the lower half rots away leaving the lower fins which look like legs." As this source concludes, "Time after time this monsterlike relic has been the cause of a sea serpent 'flap'" (Welfare and Fairley 1980, 81).

Indeed, in the case of the creature hauled up by Japanese fishermen (off the coast of New Zealand), tissue analyses were conducted by Tokyo University biochemist Dr. Shigeru Kimora. These revealed the presence of the protein elastodin, found only in sharks (Shuker 1996, 210). Other such "globsters" (as decomposed sea monsters are dubbed) turn out to be whales, oarfish, or other scientifically known creatures.

Cadborosaurus

Despite such a bleak state of affairs, an alleged sea serpent is said to appear from time to time in Cadboro Bay, on the southeast coast of British Columbia's Victoria Island. It was first reported on October 8, 1933, by a barrister, Major W.H. Langley. He was sailing in his sloop *Dorothy* about 1:30 P.M., whereupon he spied a creature "nearly eighty feet long and as wide as the average automobile." Langley said it was greenish brown and had a serrated body, "every bit as big as a whale but entirely different from a whale in many respects." His sighting was reported in the *Victoria Times* by reporter Archie Willis, and a newspaperman from the rival Victoria *Daily Colonist*, Richard L. Pocock, dubbed it "Cadborosaurus" (after its habitat, Cadboro Bay, and the Latin word for "lizard," *saurus*).

Other sightings soon followed, one on November 29, all made newsworthy by interest in reports and photos of the newly "discovered" Loch Ness Monster. Just as "Nessie" made frequent appearances in her northern Scotland home, "Caddy" became a claimed resident of the bay, and by 1950

40

some five hundred witnesses claimed to have sighted the creature (Colombo 1988, 379–380).

I can attest that Cadboro Bay is picturesque, even at night, but I suspect there is no Cadborosaurus. The many reports and accounts, I learned, "differed in details" (Colombo 1988, 380)—an indication that there may have been various creatures swimming in the waters off Victoria. As I learned in investigating lake monsters (Radford and Nickell 2006, 117–118), multiple creatures—such as otters swimming in a line—can easily be mistaken for a single long one appearing to have multiple coils or humps.

Indeed, that may explain one such Caddy sighting, at Roberts Creek, a community overlooking the Strait of Georgia (between Vancouver Island and the British Columbia mainland). It was made in 1932 by local novelist Hubert Evans (1892–1986) who saw "a series of bumps breaking the water, all in dark silhouette, and circled with ripples." He told a friend: "Sea lions. They run in a line like that sometimes." But as they watched, the profile of a head emerged which the two men estimated was extended some six feet out of the water (Colombo 1988, 369–370). However, the creature or creatures were apparently some distance away and could have been misperceived. The story was half a century old when told and related by a rather obvious romantic who gushed, "It just put the hair up on the back of your neck" (Colombo 1988, 370).

Another reported Caddy sighting (so-called, although actually occurring in the San Juan Islands chain) illustrates a similar viewing problem. Terry Graff (2006, 3) reported seeing, in 1997, "what looked like three seals in a row not thirty feet offshore," but then "realized there was only a head on the first one and the second and third were undulating humps moving up and down." I would add, "or so it seemed." Whereas one fellow eyewitness thought it a whale or seal, Graff thought it resembled Ogopogo—actually a purported Pacific Northwest lake monster (Nickell 2006)—stating, "The feeling when you see one is incredible; your mind goes into overdrive trying to classify what your eyes see and the moment you realize that it isn't classifiable is awesome!" All we can really conclude from Graff's account is that viewers were unsure of what they saw.

I got a good idea of just how difficult it can be to know exactly what you are seeing, when on board our cruise ship in Glacier Bay's Tarr Inlet, I had a creature sighting and soon thereafter spoke to a U.S. Park Service ranger about it. She told me it was probably just what I suspected—a sea otter—having actually seen otters at that place and time herself (Cahill 2006).

Two days later, while we were docked at Sitka, Alaska, I went out on a three-hour search—called Sea Otter & Wildlife Quest—aboard the double-decked excursion boat, *St. Eugene*. In addition to "Whale Rock"—a formation located just under water with waves breaking on it that is often mistaken for a whale—I saw a variety of creatures that under the right conditions could simulate a sea serpent. They included a humpback whale, a group of playful sea otters, and harbor seals basking on a little island. These mammals and others, including sea lions, represent much more likely candidates for Caddy than some imagined, hitherto unknown, leviathan.

*   *   *   *

Following publication of my investigation of cadborosaurus, I received a letter from British Columbia skeptic Steve Koerner, who related his own revealing adventure with Caddy. It was 1996 and he was on a cliff overlooking Telegraph Cove, only about half a kilometer further east from Cadboro Bay. He spotted what appeared to be a great sea serpent coming toward the beach. He watched intently, sure in the knowledge that he was not imagining what he saw—and neither hallucinating the incredible sight. Then he was astonished to see one otter climb out, "soon followed," he went on to say, "by the rest of the family" (Koerner in Nickell 2015).

References

Cahill, Adrianna. 2006. Interview by Joe Nickell, May 30.

Colombo, John Robert. 1988. *Mysterious Canada*. Toronto: Doubleday Canada Limited.

Graff, Terry. 2006. Quoted in "Eyewitness comes forward with possible Caddy report," *BCSCC Quarterly* No. 60 (publ. of British Columbia Scientific Cryptozoology Club, winter, 3).

Nickell, Joe. 2006. The Ogopogo expedition. In Radford and Nickell 2006, 111–120.

———. 2015. Monster Lookalikes: Reflections of a Paranatural Naturalist. *Skeptical Inquirer* 39: 3 (May/June), 12–15.

Radford, Benjamin, and Joe Nickell. 2006. *Lake Monster Mysteries: Investigating the World's Most Elusive Creatures*. Lexington, Kentucky: University Press of Kentucky.

Shuker, Karl P.N. 1996. *The Unexplained: An Illustrated Guide to the World's Natural and Paranormal Mysteries*. North Dighton, Mass.: JG Press.

Welfare, Simon, and John Fairley. 1980. *Arthur C. Clarke's Mysterious World*. New York: A & W Visual Library.

The Silly Ness Monster

Located in the Scottish Highlands, and stretching for almost twenty-three miles, is a narrow, trench-like basin, 227 meters in depth at its deepest point, holding the greatest volume of fresh water in all of Britain. This is the world's most famous lake, Loch Ness, home to the world's most famous—and least terrifying—legendary lake monster. "Nessie" is best known from a 1934 photograph, showing it as a long-necked, small-headed, plesiosaur-like creature but in fact proving to have been a hoax made by photographing a small model (Nickell 1995, 242).

Now, it is not true that Loch Ness is unique among Scottish lakes, or even nearly so, in having a resident monster. Whereas one writer claims there are only two others in Scotland, Loch Morar and Loch Tay, Ronald Binns (1984, 183) observes that there are "not three but dozens of 'Lochs na Beiste,'" including Loch Lomond. Admittedly that lake (which I also visited) has little monster tradition, consistent with the statement of a gentleman we encountered there; he announced emphatically that there was "no monster." He is probably correct, although some locals speak of "Lomonda" who never caught on ("Balloch and Loch Lomond" 2006). Loch Ness is of course a different story.

5.1. A composite iconographic portrait of Nessie—a metamorphosing, contortionistic, chameleon-like creature, or the product of various monster lookalikes and misperceptions. (Illustration by Joe Nickell)

Water Kelpies

The earliest recorded encounter with Nessie is told in *The Life of St. Columba*. Supposedly Columba (a Christian missionary, 521–597 CE) sent a man to swim across the River Ness to retrieve a boat, when he was approached by the monster. The future saint quickly "formed the sign of the cross in the empty air" and commanded the creature to go back, whereupon it fled. That is a saint's legend—one of a genre of narratives that are intended to demonstrate the power of God and that are notoriously incredible. Moreover, the account was written circa 700 CE, well over a century after the events they purport to describe (Shine 2006, 6; Coulson 1958, 134).

Many Scottish legends tell of kelpies or water-horses, supernatural water creatures that prey on humans. "Their trick," states Stuart McHardy, in his *Tales of Loch Ness* (2009, 52), "was to appear as a wandering horse and induce one or more humans to climb on their backs. Then they would make straight for their watery home and dive in, their supernatural powers keeping the humans on their backs as if glued." (See also Leach 1984, 573.)

Today, kelpies are relegated to the distant past, but as recently as July 1, 1853, the *Inverness Courier* reported the sighting of two strange animals.

Some thought a single sea serpent was coiled at the surface, while others suspected a pair of whales or large seals. (The sighting was at Lochend, a village near the sea-connected northeastern end of the lake, where such animals might make their way.) However, on seeing the creatures, a "venerable patriarch" threw down his gun, shouting, "God protect us; they are the Water Horses." He was partly right: they turned out to be ponies that had wandered from an estate more than a mile distant (Shine 2006, 6).

Looking back, it is easy to see that any of several creatures might have given rise to water-horse sightings—in addition, of course, to actual horses and ponies! Other possibilities are the occasional swimming roe deer, which lack true antlers but instead have short double-pronged horns (Binns 1984, 191– 93), and the sturgeon, with its long muzzle. Says naturalist Adrian Shine (2006, 28), "If ever a fish had a horse's head it would certainly be the Sturgeon."

Rounding Up Suspects

On our four-day excursion to Loch Ness, with British skeptic and investigator Hayley Stevens and her father, we found the locals to be quite playful regarding the "monster." For example, at the Glen Ord distillery where I told a tour guide we had come to the area to hunt Nessie, the woman quipped that given our quarry we were quite right to make a distillery our "first stop." In booking passage on a sonar-equipped boat to tour the loch, I asked if this were where we got our hunting license for the monster, to which a ticket seller dead-panned that we would have better luck hunting Nessie on land—meaning where the toy creatures are sold. And on the boat itself, when I wondered aloud whether rifles would be available to Nessie hunters, a boat hand quickly answered that, no, harpoons would be issued "when needed."

5.2. Investigator Hayley Stevens monitors sonar scans aboard an excursion boat on Loch Ness. (Photo by Joe Nickell)

More seriously, we made several inquiries. An information officer in Inverness told us she had seen seals in the River Ness, which passes through the city, linking the loch with the North Sea. Seals are among the monster lookalikes that are endemic either to the loch or its environs. Although the sturgeon is now almost extinct in Scottish waters, one could have occasionally entered the loch to become, say, the "huge fish" of 1882. Additional potential "monsters" are salmon and other fish, swimming deer, long-necked water birds, bobbing logs, boat wakes, wind slicks, and more (Shine 2006).

Then there is the large European otter (*Lutra lutra*). Otters resemble seals, and both are able to appear on land. However, otters also appear in rivers and lakes that are not accessible to seals. The otter's long neck and undulating movements can give it a monsterlike look (Binns 1984, 186–91). This is especially so when two or more are swimming in line, as they do because they enjoy "chasing each other" and "following the leader" (Godin 1983). They can thus appear as a single long-necked, multi-humped creature. This illusion was observed as early as 1930 in the Scottish river Clyde, and it has since been confirmed elsewhere (Gould 1934, 115–17; Nickell 2007, 6–7). (It is the basis for a delightful children's tale titled "Otterly Impossible." The story is in *Fairy Tales Fairly Told* by Barbara Mervine, with illustrations by Noah Whippie, a forward by James Randi, and inspiration for the Ness tale credited to yours truly.)

Iconography

The iconography (or study of images) of the Loch Ness Monster is illuminating. From St. Columba's "Water Beast"—with "gaping mouth and with great roaring"—and Scottish folklore's horse-headed kelpie, to the plesiosaur-like creature of the 1934 hoax photo, the monster has continued to evolve. It has since been described as six to 125 feet in length; shaped like a great eel of a creature with up to nine humps; colored silver, gray, blue-black, black, or brown; and endowed—or not—with flippers, fins, mane, horns, or tusks (Binns 1984; Gould 1934).

No doubt the extremely diverse descriptions are explained by the differences in Nessie lookalikes—as well as the vagaries of eyewitness evidence. The infamous 1934 hoaxed photo obviously did much to influence the developing portrait, although paleontologists have since postulated that plesiosaurs were unable to raise their long necks (Shine 2006, 11). In Figure 5.1, I have compressed the iconography of Nessie into a single humorous image that, I trust, makes an effective point.

Nessie Hunters

At the Loch Ness Exhibition Centre in Drumnadrochit, we received the generous attention of Adrian Shine, the venerable naturalist and Loch Ness expert, who has been a Nessie hunter since 1973. The Centre is brilliantly

48

conceived, a wonderful science museum disguised as a Nessie exhibit. It teaches continental drift, evolution, perception, and a number of other scientific matters while skeptically—but tactfully—discussing the famous monster. Shine's book *Loch Ness* supplements the tour with fascinating data and pictures—including one that remarkably recreates the previously mentioned 1934 hoaxed photo (2006, 11).

We also arranged to meet Steve Feltham, who has a little trailer on the loch at the village of Dores (near the inn, which serves delicious haggis). The world's only full-time Loch Ness Monster hunter, Feltham is the most interesting Ness creature I encountered. Having lived on the Loch since 1991, he supports himself by selling whimsical little Nessies (made of polymer clay and baked in his oven), along with driftwood art and watercolor paintings. An intelligent, thoughtful man, Feltham is no knee-jerk believer, having many times explained a tourist's sighting as, for instance, a boat wake's delayed arrival. Most recently he branded a photo of Nessie a hoax (involving a fake fiberglass hump used to film a National Geographic documentary ["Loch Ness" 2012]). He told us that even if there were no monster legend, he would still have his beachcomber life—"only in a warmer place!"

A highlight of the trip, and of my years of work as a skeptical cryptozoologist (with many appearances as such on TV series like National Geographic's *Is It Real?*, History Channel's *MonsterQuest*, and Animal Planet's *Lost Tapes Series*), was my time spent surveying the loch's waters, especially near Urquhart Castle, the location of many sightings. Our boat crossed the loch from the opposite shore to the castle and back, while Hayley and I alternated between searching the waters and monitoring sonar scans (see Figure 2). Alas, as so often is the case, Nessie was a no-show.

References

Balloch and Loch Lomond. 2006. Online at www.fermentmagazine.org/Scotland2/scot4. html; accessed September 11, 2012.

Binns, Ronald. 1984. *The Loch Ness Mystery Solved*. Amherst, NY: Prometheus Books.

Coulson, John, ed. 1958. *The Saints: A Concise Biographical Dictionary*. New York: Hawthorne Books.

Godin, Alfred J. 1983. *Wild Mammals of New England*. Chester, CT: Globe Pequot Press.

Gould, Rupert T. 1934. *The Loch Ness Monster and Others*. Reprinted Secaucus, NJ: Citadel Press, 1976.

Leach, Maria, ed. 1984. *Funk & Wagnall's Standard Dictionary of Folklore, Mythology and Legend*. New York: Harper & Row.

Loch Ness Monster photograph branded a hoax by Nessie enthusiast. 2012. Online at http://news.stv.tv/highlands-islands/186792-loch-ness-monster-photograph-branded-a-hoax-by-nessie-enthusiast/; accessed September 23, 2012.

McHardy, Stuart. 2009. *Tales of Loch Ness.* Edinburg: Luath Press.

Nickell, Joe. 1995. *Entities*. Amherst, NY: Prometheus Books.

———. 2007. *Adventures in Paranormal Investigation*. Lexington: University Press of Kentucky.

Shine, Adrian. 2006. *Loch Ness*. Drumnadrochit, Scotland: Loch Ness Project.

Quest for the Giant Eel

On a six-day trip to the Canadian province of Newfoundland and Labrador (in part for a television documentary), I encountered some very large creatures: several moose (the largest land mammal of the region), to whom I gave the right of way in return for their photos; a stuffed polar bear (towering upright almost nine feet tall), which had ambled into St. Anthony one spring; and, from a circus truck that overturned ahead of me on the Viking Trail, two camels and a sweet Asian elephant named Limba.

I did not encounter humpback whales, although I took an excursion boat out in very rough water to see great icebergs making their way south from Greenland. (I had better luck with humpbacks on an Alaska excursion [Nickell 2007a].) Neither did I catch a glimpse of another leviathan that occasionally haunts the region's coastal waters: the subject of chapter 1, the giant squid, known at lengths up to sixty feet and the subject of numerous hair-raising adventures (Fitzgerald 2006, 50–71). (For our book *Lake Monster Mysteries*, Benjamin Radford [2006, 5] photographed the world's best-preserved specimen at a museum in St. John's.)

Figure 6.1. For the TV series *MonsterQuest*, the author visited Crescent Lake, Newfoundland, where "Cressie" is reported to lurk—possibly as a giant eel.

What I was really searching for—having been brought to the village of Robert's Arm by a television crew for the History Channel's popular series, *MonsterQuest* (which later aired on September 17, 2008)—was a legendary lake monster said to inhabit the cold, deep, blue waters of Crescent Lake. It has been dubbed "Cressie," and the village's welcoming signboard proclaims it "The 'Loch Ness' of Newfoundland!"

'Cressie'

Sightings of a "monster" in the lake date back to the turn of the last century when a resident known as "Grandmother Anthony" spied a giant serpentine creature while she was picking berries. From the 1940s to the present, there have been a dozen or so sightings, although without photographs to date. Most descriptions are of a dark, eel-like creature, up to twenty-five or more feet long (Bragg 1995; Radford and Nickell 2006, 89–95).

Its locomotion is typically described as "rolling" or "undulating" (Bragg 1995); indeed, "when the head was up, the back was down" (Colbourne 2008). Consequently, the contortions of the elongated creature seemingly produced "humps" (Short 2008; see figure 6.1).

53

A typical sighting occurred in 1991, when retired school teacher Fred Parsons (an engaging man whom I met in Robert's Arm) saw a creature surface while crossing the lake. It was dark brown, swimming in an undulating fashion, and, Parsons estimated, over twenty feet long (Bragg 1995; see also Radford and Nickell 2006, 92–93). Of course eyewitness testimony can be unreliable. An experiment I conducted for *MonsterQuest*, using a log of known length that we towed and anchored at a mid-lake position, demonstrated that people viewing something from a distance can easily overestimate its size by forty percent or greater.[1]

There are other reasons to be skeptical of a monster in Crescent Lake, one of which is that a single creature could neither live for centuries nor reproduce itself. A breeding herd of several individuals would be required for the species to continue propagating over time. But then where is a single floating or beached carcass? It is true that the lake is connected to the Atlantic Ocean, scarcely two miles distant, by Tommy's Arm Brook. However as Bragg (1995) concedes, no great creature has ever been seen navigating the outlet.

Giant Eel?

Because "Cressie" is often likened to a giant eel (Bragg 1995; Eberhart 2002, I:114; *Monster* 2008), someone gave it the quasi-scientific name *Cressiteras anguilloida* (Eberhart 2002, I:114). Actually, this is unlikely as a scientific name that might be bestowed—if a giant-eel specimen were verified. Eels (a group of fishes having snakelike bodies and lacking pelvic fins) are of the order *Anguilliformes*, and true eels comprise the family *Anguillidae*. The American eel, for example, is *Anguilla rostrata* (Collins 1959, 475). Related eels include the marine conger eels (*Conger oceanicus*), which attain a length of six to nine feet, and the morays of tropical reefs. The Pacific moray (*Thyrsoidea macrurus*), up to a foot longer, "is probably the largest known species" (*Colliers Encyclopedia* 1993, s.v. "Moray").

Now, while Crescent Lake does reportedly host freshwater American eels, these are normally under five feet long. Divers from the Royal Canadian

Mounted Police (RCMP), who allegedly surfaced on the lake with "descriptions of giant eels as thick as a man's thigh" (Bragg 1995), probably encountered a different creature—if indeed, the incident actually happened: The RCMP could not confirm the occurrence to *MonsterQuest*. Indeed, whatever Cressie is, it is clearly not a giant eel. The eyewitness descriptions of a giant creature, swimming on the surface of the water and moving in an up-and-down fashion, are completely wrong for an eel.

Eels, in fact, are bottom-dwelling creatures ("Freshwater" 2008a, 2008b; "Eel" 2008), and their locomotion, while wavelike, is actually from side-to-side, as I confirmed by studying them at Aquarium Niagara in Niagara Falls, New York (where I am a member and once served as "Animal Trainer for a Day"). For my *MonsterQuest* research, the aquarium's exhibits supervisor, Dan Arcara, graciously allowed me to study an American eel and a moray eel, gently prodding the latter from its den with a pole so I could document on videotape its sideways-oscillating swimming style.

Moreover, the sightings of Cressie invariably occur during daytime, whereas the common freshwater eel "is nocturnal in its habits, sleeping or lying in the mud during the day" ("Freshwater" 2008a).

Cressie Lookalike

There is, in fact, an actual creature that is dark-colored, swims both under water and at the surface—where its wake can make it appear much longer—and moves in an undulating (rising and falling) manner. Its scientific name is *Lontra canadensis*,[2] the northern river otter (Nickell 2007c).

In addition, multiple otters swimming in a line can give the effect of a single giant serpentine creature slithering with an up-and-down movement through water. This effect was observed as early as 1930 by a marine biologist (Gould 1934, 115–116) and has since been documented many times (e.g., Nickell 2007b). Newfoundland is shown (by the *National Audubon Society Field Guide to Mammals* [Whitaker 1996, 782–785]) to be a definite habitat for the northern river otter. (See figure 6.2.)

Figure 6.2. Giant eel or otter lookalike?

I have been accused of seeming to suggest this effect as a solution to all lake monster reports (Coleman 2007), but in fact that grossly mischaracterizes my position. In *Lake Monster Mysteries*, I acknowledged other lake-monster imitators, including fish (such as sturgeon and gar), long-necked birds, wind slicks, boat wakes, and logs (which may be propelled from the lake bottom by methane gas produced by decomposition [Monk 2004]). Swimming mammals like deer and beaver have also been mistaken for lake monsters. For instance, during the filming of the *MonsterQuest* program, a mysterious and seemingly lengthy creature swimming under the surface of the lake created a brief sensation but proved to be a beaver.

I apply otters as a solution to *some* mystery sightings, according to the principle of Occam's razor (that the simplest credible solution, the one making the fewest assumptions, is to be preferred). When a sighting could most credibly be explained as one or more otters, like some of the Cressie sightings, then that is necessarily the preferred hypothesis. Other sightings may be attributed to other causes. However, should Cressie surface in a more credible form, I would certainly be willing to reopen the case.

Notes

1. This was conducted on Saturday, June 14, 2008. Two of the three participants—Bradley Rideout and Effie Colbourne—had reported seeing "Cressie." Brad estimated the 14.25-foot log at 18 feet, Effie at 20 (although first saying "20 to 30"), and the other participant at 20 feet.

2. Formerly *Lutra canadensis*.

References

Bragg, R.A. 1995. Have you seen Cressie? In Wanda Jackman, Bonnie Warr, and Russell A. Bragg, *Remembrances of Robert's Arm*. Corner Brook, Newfoundland: Western Star Publishers, 14.

Colbourne, Effie. 2008. Interview for *MonsterQuest* (*Monster* 2008).

Coleman, Loren. 2007. Otter nonsense. Available online at www.cryptomundocom/cryptozoo-news/otter-nonsense; accessed June 6.

Eberhart, George M. 2002. *Mysterious Creatures: A Guide to Cryptozoology* (in two vols.). Santa Barbara, Calif.: ABC-CL10.

Eel. 2008. From Wikipedia, available online at http://en.wikipedia.org/wiki/Eel; accessed August 20, 2008.

Fitzgerald, Jack. 2006. *Newfoundland Adventures: In Air, on Land, at Sea*. St. John's, Newfoundland and Labrador: Creative Publishers.

Freshwater eels. 2008a. Available online at http://gamefishingguide.com/freshwater-eels.html; accessed August 8, 2008.

Freshwater vs. saltwater moray eels revisited. 2008b. Available online at http://Saltaquarium.about.com/cs/eelcare/a/aa090501.html; accessed August 20, 2008.

Gould, Rupert T. 1934. *The Loch Ness Monster*; reprinted Secaucus, N.J.: Citadel Press, 1976.

Monk, Jerry. 2004. Letter to the editor. *Fortean Times* 185 (July): 76.

MonsterQuest eyewitnesses. 2008. Transcript of preliminary interviews for *MonsterQuest*, provided to author September 6.

Nickell, Joe. 2007a. Mysterious entities of the Pacific Northwest, part I. *Skeptical Inquirer* 31:1 (January/February), 20–22.

———. 2007b. Lake monster lookalikes. *Skeptical Briefs*. June, 6–7.

———. 2007c. The Loch Ness critter. *Skeptical Inquirer* 31:5 (September/October), 15–16.

Radford, Benjamin, and Joe Nickell. 2006. *Lake Monster Mysteries: Investigating the World's Most Elusive Creatures*. Lexington, Ky.: The University Press of Kentucky.

Rideout, Bradley. 2008. Interview for *MonsterQuest* (*Monster 2008*).

Short, Vivian. 2008. Interview for *MonsterQuest* (*Monster 2008*).

Whitaker, John O., Jr. 1996. *National Audubon Society Field Guide to North American Mammals*. New York: Alfred A. Knopf.

Lake Cryptid Lookalikes

As a cryptozoologist—albeit a skeptical one—I have long been on the track of fabled creatures, culturally and historically (Nickell 1995; 2006) as well as investigatively. Among my quarry have been legendary leviathans like those supposed to inhabit lakes Champlain (New York and Vermont), Memphremagog (Vermont and Quebec), Utopia (New Brunswick), Okanagan (British Columbia), Simcoe (Ontario), Silver Lake (New York) and others.

Indeed, I co-authored a major study of the phenomenon, *Lake Monster Mysteries* (Radford and Nickell 2006), which has received very good reviews—even from "believers." For example, the newsletter of the British Columbia Scientific Cryptozoology Club (of which we are both members) spoke of "the incredible amount of investigative work" we had "put into the field of cryptid investigations," noting that we "have done more fieldwork than many cryptozoology enthusiasts and have been very diligent about it" (Kirk 2006).

The reviewer—respected cryptozoologist John Kirk, author of *In the Domain of the Lake Monsters* (1998)—did fault us a bit, finding that my co-author was somewhat too dismissive of eyewitness reliability, and that I had perhaps "too broadly applied" a particular explanation for sightings of multi-humped creatures.

Actually, I feel that if we erred it was on the side of being too open-minded. An appendix, "Eyewitness (Un)Reliability," simply demonstrated the fact that eyewitnesses are often mistaken. If further evidence is needed, consider a case that transpired in Rotterdam in 1978. A small panda had escaped from a zoo, whereupon officials had issued a media alert. Soon panda sightings—around one hundred in all—were reported across the Netherlands. However, a single animal could not have been in so many places in so short a time; in fact, no one had seen the panda, because it had been killed by a train when it reached railroad tracks near the zoo. How do we explain the many false sightings? The answer is, people's anticipations led them to misinterpret what they had actually seen—a dog or some wild creature—as the escaped panda. (The publicity generated by the case may even have sparked some hoax calls [Nickell 1995, 43].) If such misperceptions could happen with pandas, surely they could also occur with aquatic cryptids.

Which brings me to Kirk's criticism (mild criticism, to be sure) of my hypothesis regarding certain sightings. As suggested in the previous chapter, I had discovered that there was a notable lake-monster lookalike in the vicinity of many reported encounters. Consider, for example, the experience of a senior wildlife technician with New York's Department of Environmental Conservation, Jon Kopp. As he explained to me, it had been dark and he was in a duck blind on a lake in Clinton County. Suddenly, he saw, heading toward him, a huge, snake-like monster swimming with a sinuous, undulating motion. As it came closer, however, Kopp realized that he saw not one creature but half a dozen—a group of otters swimming in a line diving and resurfacing to create the effect of a single, serpentine creature. "After seeing this," Kopp said, "I can understand how people can see a 'sea serpent'" (Radford and Nickell 2006, 38).

As another example, one witness to a Lake Champlain sighting of 1983 stated, "It could have been one large creature or four smaller ones," a concession underscored by the fact that the sighting was at the "mouth of Otter Creek"—actually Vermont's longest river, but in any case aptly named as a habitat for the northern river otter (Zarzynski 1983).

Composite drawing of the Lake Okanagan monster.

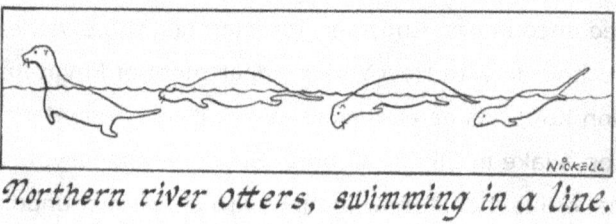

Northern river otters, swimming in a line.

Figure 7.1. Author's composite drawing of Ogopogo (top) is compared with otters swimming in a line (after Gould 1976).

That creature—*Lutra canadensis*—measures up to fifty-two inches long, and when treading water with its hind paws, can extend its head and long neck well out of the water. It thus invites comparison with the extinct plesiosaur, which is often cited as a possibility for supposed lake-monster sightings (Binns 1984, 186—191). Moreover, otters are playful and enjoy "chasing each other" and "following the leader" (Godin 1983, 173), thus being especially prone to creating the illusion of a single multi-humped creature. Figure 7.1 illustrates how this behavior can help explain, for instance, many of Lake Okanagan's fabled "Ogopogo" sightings.

Of course, otters are not responsible for all lake-monster sightings, any more than weather balloons are the only instigators of UFO reports. In fact, in *Lake Monster Mysteries*, I mentioned many possible culprits, such as sturgeon, gar, and other large fish; swimming animals like beavers; deer; long-necked birds; bobbing logs; clumps of dislodged lake-bottom debris; and additional possibilities, including wind slicks and boat wakes. Hoaxes are also possible, and there have been faked monsters on pulleys as well as phony photographs, like the celebrated Loch Ness monster photo, which was publicly revealed as a hoax in 1994 (Radford and Nickell 2006).

64

Nevertheless, in light of considerable evidence, I decided to see how well monster-inhabited lakes and rivers correlated with otter populations, and my research led to the map in figure 7.2. For this, I relied on the list of "Lake Monsters of the World" given by Kirk (1998, 293—303), and for otter distribution, I consulted the *National Audubon Society Field Guide to North American Mammals* (Whitaker 1996).

For practical reasons,[1] I limited my scope to North America, but it should be noted that otters (any of various aquatic carnivores belonging to the weasel family) are "found on all continents except Australia" (*Webster's* 1997). At Loch Ness, for example, the European otter (*Lutra lutra*) is an often-cited potential lookalike for Nessie (Binns 1984, 186—191).

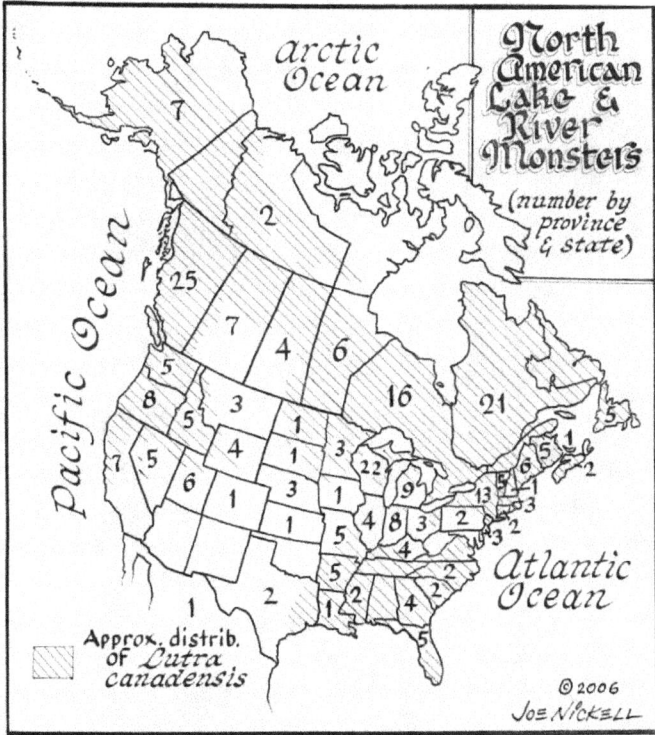

Figure 7.2. This map shows the numbers of monster-inhabited lakes and rivers correlated with the distribution of otter populations.

Clearly, there is a strong correlation as expected. Note that the large number of aquatic-cryptid sites in British Columbia (25), Ontario (16), and

Quebec (21), are in prime river-otter territory. Of course it can be argued that lakes themselves are in abundance there, and that, conversely, where water is relatively scarce (e.g., the southwestern United States), naturally otters are also scarce. But I rather think that that helps prove my point: wherever there are lakes and rivers with "monsters"—especially of the long-necked multi-humped variety—otters are usually known to the area. They thus become viable candidates for the apparent misperception—a variation on the old if-it-looks-like-a-duck adage.

However, we must continue to investigate claims on a case-by-case basis. Our aim must be to solve lake monster mysteries, not to foster or dismiss them.

Note

1. Including all the monster-inhabited lakes of the world on a simple map would have resulted in a large illustration with too much detail to be reasonably publishable in reduced size in a magazine or book.

References

Binns, Ronald. 1984. *The Loch Ness Mystery Solved*. Buffalo, N.Y.: Prometheus Books.

Godin, Alfred J. 1983. *Wild Mammals of New England*. Chester, Conn.: Globe Pequot Press.

Kirk, John. 1998. *In the Domain of the Lake Monsters: The Search for Denizens of the Deep*. Toronto: Key Porter Books.

———. 2006. Cryptozoological publications, media and film reviews. *BCSCC Quarterly*, Spring, 11.

Nickell, Joe. 1995. *Entities: Angeles, Spirits, Demons, and Other Alien Beings*. Lexington, Ky.: University Press of Kentucky.

———. 2005. *Secrets of the Sideshows*. Lexington, Ky.: University Press of Kentucky.

Radford, Benjamin, and Joe Nickell. 2006. *Lake Monster Mysteries: Investigating the World's Most Elusive Creatures*. Lexington, Ky.: University Press of Kentucky.

Webster's New Universal Encyclopedia. 1997. New York: Barnes & Noble, s.v. "otter."

Whitaker, Jon O., Jr. 1996. *National Audubon Society Field Guide to North American Mammals*. New York: Alfred A. Knopf.

Zarzynski, Joseph. 1983. LCPI Work at Lake Champlain. *1983.*
Cryptozoology 1: 73—77.

Arkansas's White River Monster: Very Real, But What Was It?

Arkansas's White River holds claim to a "monster" that has been said to have appeared intermittently for a century or more. Its reality is defended by cryptozoologists and skeptics alike (I am both), but what *is* the huge aquatic creature that has proven so elusive? The late cryptozoologist Roy Mackal (1980) thought he knew, but there were problems with his theory. Can we finally solve the mystery?

Monster and Lair

Sightings have been claimed from 1912 and 1917, among other years, but it was July 1937 that confirmed the existence of a great river creature—albeit one that is variously described. At some point, named for its lair, it was dubbed "Whitey," although almost everyone agreed its color was gray. Some added that its skin was "crusty" or "peeling."

Its size was not so certain, with estimates ranging from at least twelve feet to sixty-five feet or more—although Mackal wisely cautioned (1980, 205):

> I ask the reader to take my word for the observation that untrained observers estimating size of unknown objects in water range from exactly right to threefold or fivefold too great. This is especially true when the observer has no object of known size with which to make a comparison. It is generally not appreciated to what extent we rely on comparison to make size estimates.

The creature's weight was also variable—1,000 pounds or more. Although the head was almost never seen, one eyewitness reported glimpsing it and thought it had a protruding horn (Mackal 1980, 207; Cox 2011).

The 1937 sightings placed the creature at a portion of the White River—a major tributary of the Mississippi—just below the city of Newport. A farmer named Bramlett Bateman signed an affidavit as to his observations—which began at about one in the afternoon on July 1 and lasted for five minutes. He subsequently saw it several times over more than two months and believed it to be "about 12 feet long and 4 or 5 feet wide." Three other affidavits were produced—one by a deputy sheriff—and Bateman said he knew of two dozen others who could similarly attest to the unusual creature. Resulting newspaper accounts catapulted the monster of White River "into national prominence" (Mackal 1980, 200).

During the flap, a woman named Ethel Smith of Little Rock stated she had seen the creature thirteen years earlier—i.e., in 1924—while vacationing with her husband and children. Elements of her description tallied with those of others: "It was making a loud blowing noise but never did show its head or tail. It was a terrible-looking thing with dingy gray crusted hide. It frightened me badly." A local fisherman said he had also seen the creature previously at the same site around 1915, and sightings were also reported in 1971 and 1972.[1] As was common, the creature appeared following a widening ring of bubbles and thrashed about for five minutes or more before resubmerging (Mackal 1980, 204). It left in its wake open-mouthed eyewitnesses.

Mackal's Identification

Roy Mackal (1925–2013), a biologist as well as a cryptozoologist, featured the White River Monster in his seminal cryptozoological book, *Searching for Hidden Animals* (1980, 197–208). He brought a great deal of expertise and good sense to the subject.

Mackal was persuaded that those who encountered the White River creature offered useful testimony: "Except for size there is a remarkable consistency among the different witnesses over a long time. There can be no doubt that a real animal or animals had been observed" (Mackal 1980, 205). Mackal collected and summarized the applicable zoological details to produce a sort of composite picture of the creature, as we have seen.

He concluded:

> The White River case is a clear-cut instance of a known aquatic animal observed outside of its normal habitat or range and therefore unidentified by the observers unfamiliar with the type. The animal in question clearly was a large male elephant seal, either *Mirounga leonina* (southern species) or *Mirounga angustirostris* (northern species).

His identification has convinced many, but there is a serious problem.

Ranging Afar?

The creatures (it's unlikely to be the same individual given such intervals) would have had to enter the river from the Mississippi, and to have connected with it from the Gulf of Mexico. However, it is more than difficult to hypothesize, in the gulf waters, either species of the elephant seal. The Northern species ranges over the Pacific coast of northern Mexico, the United States, and Canada (Whitaker 1996, 742–744), while the Southern elephant seal migrates from Antarctic and sub-Antarctic waters north to Argentina, New Zealand, and South Africa, and even wandering individuals are not found above the equator ("Ocean" 2018).

Another plausible aquatic mammal is the hooded seal. Although it is smaller than the Northern elephant seal (at up to ten feet versus thirteen), it could fit descriptions of "Whitey" pretty well. Although the hooded seal's habitat is the Arctic and North Atlantic, it is very migratory, and individuals have "strayed as far south as Florida" (Whitaker 1996, 739–741). Nevertheless, I am unaware of one ever having reached the Gulf of Mexico, let alone to have then traveled far up the Mississippi and to have done so repeatedly.

One of the seals, the Caribbean or West Indian monk seal, did once populate the very waters of the gulf. Unfortunately, it is now extinct (Whitaker 1996, 741–742).

I believe there is yet another possibility, and it is not only found in the Gulf of Mexico but known to have swum hundreds of miles up the Mississippi!

The Best Suspect

I am referring to the Florida manatee (a subspecies of the West Indian manatee).[2] It is a "massive" aquatic mammal with a minimum length of about thirteen feet (although the largest recorded was fifteen feet). It may weigh as much as about 3,500 pounds. Like the White River monster, it has gray, smooth skin that can appear mottled due to barnacle-like crusts of algae or to common injuries from boat propellers (Whitaker 1996, 807–808; "Florida" 2018; "West Indian" 2018).

It has no "horn," but whereas Mackal postulated that once-only descriptor was due to the elephant seal's proboscis (short trunk), I suggest it was one of the manatee's front flipper-like legs, seen beside its head when it rolled over in the water. These appendages have three nails at their end and would seem capable of leaving on shore the fourteen-by-eight-inch, three-toed tracks attributed to the "monster," which also flattened grass and saplings (Mackal 1980, 203, 205, 207). Manatees do crawl onto shores to graze on plants as part of their herbivore lifestyle. And just like the monster, the manatee also makes "blowing" noises ("Manatee" 2014). Again, it basks on the water, rolls, dives, and so on ("West Indian" 2018).

Significantly, the manatee is adapted to both fresh and salt water and so is found in rivers and in the Gulf of Mexico as well as the Atlantic Ocean. It ranges as far west as Texas and as far north as Massachusetts. In fact, in 2006 one traveled some 720 miles up the Mississippi River to enter the Wolf River near Memphis. It was eventually found dead on the banks of McKellar Lake, a slackwater lake south of that city ("Manatee" 2006). That animal's journey shows a manatee to be a very real possibility as the White River creature.

The Memphis manatee died in October and was thought to have

succumbed to the cold. However, when a manatee is found north of Florida, it is as "mainly a summer immigrant" (Whitaker 1996, 807). That is consistent with the fact that the White River monster was observed mostly during the summer months: July 1924; June to early September 1937; June, July, and August 1971; and June 1972.

All things considered, the Florida manatee surely represents the preferred hypothesis in the case, which I believe we may now mark closed.

Notes

1. One witness in 1971 made a Polaroid snap-shot of the supposed creature, which "appeared to have a spiny backbone that stretched for 30 or more feet" (Mackal 1980, 202). In my opinion this looks unnatural and more like a series of dashes made by a retoucher using a pen.

2. I learned a bit about manatees in 2001 when I was in Apalachicola National Forest in the Florida panhandle, which is also supposedly home to a Bigfoot-like creature called the Florida Skunk Ape.

References

Cox, Dale. 2011. A River Monster in Arkansas? Available online
at
http://www.exploresouthernhistory.com/whiteriver1.htm
l; accessed January 29, 2018.

Florida manatees. 2018. *Marine Mammals of the Gulf of Mexico*,
brochure of the Institute for Marine Mammal Studies.

Mackal, Roy P. 1980. *Searching for Hidden Animals*. Reprinted,
London: Cadogan Books, 1983, 197–208.

Manatee found dead on banks of McKellar Lake. 2006. Available
online at
http://www.wmcactionnews5.com/story/5798366/manat
ee-found-dead-on-banks-of-mckellar-lake; accessed
January 24, 2018.

Manatee Making Strange Noise: Blowing Raspberries. 2014.
Available online at
https://www.youtube.com/watch?v=8SZ-eJ1Ub71;
accessed January 25, 2018.

"Ocean Treasures" Memorial Library. 2018. Available online at
http://otlibrary.com/ southern-elephant-seal/; accessed
January 26, 2018.

West Indian manatee. 2018. Available online at
https://en.wikipedia.org/wiki/West_Indian_
manatee#Description; accessed January 19, 2018.

Whitaker, John O. 1996. *National Audubon Society Field Guide to North America*, revised edition. New York: Alfred A. Knopf.

Gigantic Catfish: Investigating a Whopper

For its fourth season, the popular television show *Monster Fish*, on the cable channel National Geographic Wild, asked for my opinion of an old photograph depicting a humongous catfish—one estimated to weigh between 500 and 800 pounds. Was the photo authentic? I flew to Chattanooga to give my opinion for a segment of the episode "Giant Catfish" that aired July 5, 2013. Here is a more detailed presentation.

A Fish Tale

There are different versions—folklore at work—regarding the origins of both the giant fish and the old photo. Some say the picture is genuine, while others insist that it is not.

Figure 9.1. Questioned photograph: real or fake?

For example, some accounts hold that the fish was caught in the Tennessee River at the Cerro Gordo community in Hardin County, Tennessee, in 1914. Some say it was landed by the late Joe B. Pitts, the proprietor of Harbour-Pitts Company's general store, while others insist it was actually taken by Green Bailey, a local fisherman who caught it on a trotline. Another story holds that the giant fish was captured after it was trapped in shallow water during a dry spell; one local historian thinks the picture may date from the 1940s; and so on (Wilson 2003; Cagle 2010).

Photo Analysis

A copy of the original photograph (Figure 9.1) bears (in the upper right corner) a handwritten notation, "Cerro Gordo, Ta[ken?] by Green Bailey / Apr 6th / 1914" and (in the lower left) the initials "E. F. P.," presumably those of the photographer. That the script is white is consistent with its having been directly penned, probably with India ink, on the photographic negative, a common practice. (See examples in Nickell 2005, 64, 125.)

Immediately one notices that the photo's ratio of width to height is like that of a picture postcard rather than a standard photograph. Common in the period indicated were what are known to deltiologists (postcard collectors) as "real photo" postcards: black-and-white photos that were not printed on a press but were actually developed onto photographic cardstock having a preprinted postcard back.[1] By 1902, Kodak was selling such prepared cards (Nicholson 1994, 178). Real photos were often made by local or itinerant photographers and so were typically one-of-a-kind or limited-edition prints (although some were commercially made and tended to have typeface captions rather than handwritten ones). Real photos were especially common during the "golden age" of picture postcards, 1898–1918 (Nickell 2003, 105–107; Nickell 2005, 41–43, 56; Willoughby 1992, 68–77; Nicholson 1994, 3, 13, 178).

Looking at the photo image itself, it is quite typical of the postcard genre known as "exaggerations" (Range 1980, 62–63)—a genre now recognized in folklore scholarship as a form of American folk expression representing a "visual twist on the tall tale" (Axelrod and Oster 2000, 184–185). I have a collection of these: examples include a man being attacked by a monster jackrabbit, a workman hauling a single huge bunch of grapes in a wheelbarrow, two great cabbages (or again two titanic oranges) filling a train's flatcar, a tractor pulling one gargantuan potato, and so on and on. (See Figures 9.2 and 9.3.)

But fish are perhaps the most common, no doubt because, as one folklorist observes, "Verbal lore concerning fabulous catches represents a whole category of exaggerated narrative, highly formulaic and entertaining in content, with skeptical reaction from the audience an expected part of the performance" (Brady 1996). To this *verbal* lore is added the *pictorial* variety: photos or artworks that often turn into jokelore—funny instances of the one that *didn't* get away.

Now, such photographs are typically made by photomontage techniques. The term *montage* (French for "mounting") loosely describes any means of making one picture from two or more—as by background projection, collage or "cut montage," sandwiching (of negatives), and other techniques (Nickell 2005, 120–127). The figure of the man standing atop the wagon and staring at the giant catfish does have a "different" look than other elements in the photo—it is a bit out of focus, for example—and could seem to indicate photomontaging. It could, that is, if there were not additional evidence pointing to authenticity.

Figure 9.2. "Real photo" postcard of "exaggeration" genre (photo montage technique), 1908. (Author's collection)

Genuine Photo...

As I told the star of *Monster Fish*, fish biologist Zeb Hogan, the photograph is actually genuine and unretouched. But that doesn't mean the giant catfish is the real McCoy: the *scene* the genuine photo depicts has been faked!

Joe Brownlow Pitts of Savannah, Tennessee, speaks with authority: "My daddy had a little wagon that looked like a log wagon. He put the fish—which weighed, I recall, about 85 lbs.—on it. Then my Uncle Frank [Elisha

81

Franklin Pitts (1890–1953)], who was good at photography, cut out a cardboard man that was being used in a clothing advertisement and stuck it on the wagon, along with the fish. He took the picture" (qtd. in "Giant Catfish" 2007). Another source (Cagle 2010) explains that "the wagon, perhaps a quarter or less than the size of a standard wagon, was a freight wagon used to move goods in tight quarters, such as the basement of the Harbour-Pitts Company Store and was pulled by workers, not horses."

The cutout figure—known to cameramen of the period as a "photographic statuette" (a photo mounted on a rigid base, such as cardboard or plywood, and cut with an appropriate tool such as a mat knife or jigsaw [Fraprie and Woodbury 1896, 90–96])—was, as mentioned, reportedly made from an advertisement. However, another source (Wilson 2003) cites hearsay evidence that the man in the picture was one Warren McConnell. It is possible that photographer Frank Pitts posed him and took his photograph rather than using an advertising figure.[2] But what is important is that Jay Barker, president of Tennessee's National Catfish Derby, reportedly had "a copy of another photo of the same man and fish taken from a different angle"—in which "the man is posed exactly the same as he is in the other photo" (Wilson 2003). This would appear to corroborate the use of a "photographic statuette" in staging the scene.

As to the catfish itself, it may well have been caught by Green Bailey "who worked as a gin-man [cotton-gin operator] for Habour-Pitts Company." He reportedly caught it on a trotline (which would not have been strong enough to hold a 500–800 pound fish![3]). Then he "took his catch to the gin where it could be weighed on a cotton scale" (Cagle 2010). Sources say a former bookkeeper for the Pitts store, Rilla Callens, who actually had passed the photo down to her son, agreed Bailey had caught the fish; so did Green Bailey's sister (Wilson 2003).

Figure 9.3. Another "real photo" card shows one that did not get away. (Author's collection)

Re-Creation

The evidence fully explains the picture as I conclude, and so does photo expert Tom Flynn, a CFI photographer and videographer I consulted, who has expertise in special effects. Tom suggests the staged scene was photographed with a view camera, using a wide aperture and slow-speed film. That the background, especially, is out of focus is consistent with a small object being photographed close up. This effect is often seen when miniature objects are photographed with the intention of making them look larger.

It remained to do a recreation, and for that I flew to Chattanooga. A small wagon was found (on eBay, as I recall), Zeb brought a sizeable catfish in a cooler, and the National Geographic Museum shop provided the excellent cutout picture of Zeb, much smaller than life size. As with the original Frank Pitts photo, the fact that one assumes the wagon and figure are of usual size creates an illusion in which the catfish appears to be huge indeed (Figure 9.4): larger than life—in fact much, much larger.

Figure 9.4. Re-creation of the process for the photo in Figure 1. (Photo by Joe Nickell)

Notes

1. Although the image in Figure 9.1 is halftone-screened, the half-toning extends even over damaged areas, showing that it results from copying at some later time (either by a photocopier that screens or from a printed reproduction of the card).

2. Frank Pitts may not have had the capability of making such a big enlargement.

3. I have some experience with trotlines, having often accompanied my late grandfather, Charlie Turner, as he removed catfish from his.

References

Axelrod, Alan, and Harry Oster. 2000. *The Penguin Dictionary of American Folklore*. New York: Penguin Reference.

Brady, Erika. 1996. Fishing (sport), in Brunvand 1996, 274–275.

Brunvand, Jan Harold, ed. 1996. *American Folklore: An Encyclopedia*. New York: Garland Publishing.

Cagle, David. 2010. Statement and copy of original photo from Hardin County, Tennessee, Historical Society, August; provided by National Geographic Television.

Fraprie, Frank R., and Walter E. Woodbury. 1896. *Photographic Amusements*, 10th ed. Boston: American Photographic Publishing Co., 1931.

Giant Catfish—Some of the World's Biggest Catfish. 2007. Online at http://www.oodora.com/life-stories/funny-finds/giant-catfish.html; accessed August 30, 2013.

Nicholson, Susan Brown. 1994. *The Encyclopedia of Antique Postcards*. Radnor, PA: Wallace-Homestead.

Nickell, Joe. 2003. *Pen, Ink, and Evidence: A Study of Writing and Writing Materials for the Penman, Collector, and Document Detective*. New Castle, Delaware: Oak Knoll Press.

———. 2005. *Camera Clues: A Handbook for Photographic Investigation*. Lexington: University Press of Kentucky.

Range, Thomas E. 1980. *The Book of Postcard Collecting*. New
York: Dutton.

Willoughby, Martin. 1992. *A History of Postcards*. Secaucus,
N.J.: The Wellfleet Press.

Wilson, Taylor. 2003. Something's fishy in Hardin County.
ESPNOutdoors.com (November 11); accessed July 6,
2012.

Louisiana Swamp Monster

Louisiana Honey Island Swamp Monster fakery continues to sell, while sometimes common-sense skepticism on the subject has all the currency of—well, a wooden nickel. This was brought home to me personally when I was asked by a production company to look into the latest alleged exploits of the ever-elusive creature. Unfortunately, after I enlisted the aid of Tom Flynn—CFI's resident photograph, film, and video expert—the producers left us in the lurch. It's pretty obvious why: the evidence—a brief Super-8 film of the Swampster—is so bad that any critical analysis would leave one asking why the show would be made at all.

Background

What is the Honey Island Swamp Monster? It debuted in 1974 when two local hunters—Harlan Ford and Billy Mills—emerged from the pristine backwater area (see Figure 10.1) with plaster casts of "unusual tracks." These web-footed, four-toed imprints seemed "a cross between a primate and a large alligator" (Holyfield 1999, 9). A copy cast of one track I obtained from Ford's granddaughter measures only about $9^3/_4''$—about half the size of a large bigfoot print. (See Figure 10.2.) Ford and Mills also claimed that in 1963 they saw similar tracks after encountering a seven-foot-tall creature covered with grayish hair and having large, amber eyes. The two were left without evidence when, they said, the creature ran away and a rainstorm washed away its tracks.

Figure 10.1. Louisiana's pristine Honey Island Swamp is the alleged habitat of a manlike monster.

Since I investigated the case in 2000, visiting the swamp with a Cajun guide, there have been significant developments. There was the discovery of one of a pair of "shoes" for making swamp-monster tracks, buried in mud near Harlan Ford's former hunting camp (Nickell 2011, 219). This is prima facie evidence of hoaxing. The other was a reel of old Super-8 film that Ford's widow found among his belongings long after his death in 1980 (Holyfield 2007).

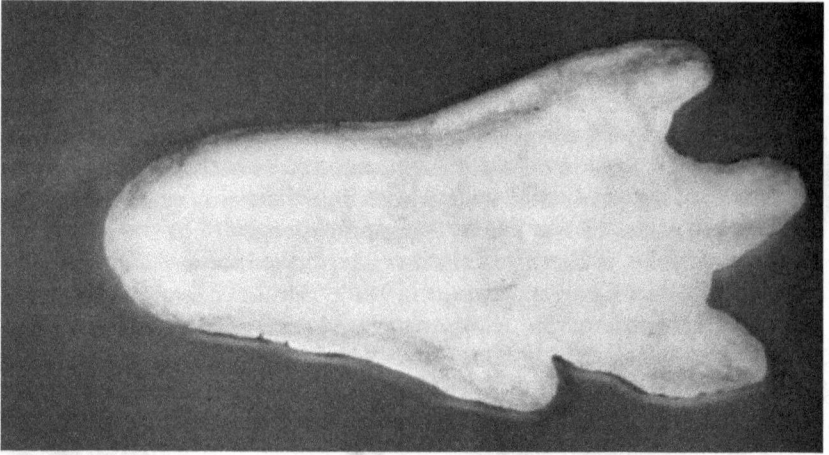

Figure 10.2. Plaster cast of alleged Honey Island Swamp monster track.

Ford's film seems to have been made after 1974 (the latest year cited in an undated, unpublished article by him that fails to mention any such film) and of course 1980 (the year of his death). His granddaughter, Dana Holyfield-Evans (2014, 36), wonders why he never revealed the film's existence, but it may have been because it looks little better than a joke. (See figure 10.3.)

Figure 10.3. Frames from an old Super-8 film show the Louisiana Honey Island Swamp Monster—or *Bigsuit*—striding through the woods. (Thanks to Cody Hashman for the screen capture.)

Analysis

Tom Flynn and I studied the film frame by frame, concluding that the production was suspicious in many ways. For example, the amateur filmmaker was able to record the introductory trip into the swamp, a view of the tree blind from which the creature was filmed, and, with only a couple of seconds' wait (!), a 7-second shot of the elusive creature—all within 2 minutes and 10 seconds of the available 3 minutes and 30 seconds allotted for a roll of Super-8 film. Also, some of the ending and possibly the beginning of the film was suspiciously missing.

The creature appears not to be Bigfoot but rather Bigsuit. Its brief appearance on camera and its being hidden from clear view mean that a very cheap suit could have been employed. The idea no doubt derived from the infamous Roger Patterson film of a female sasquatch at Bluff Creek, California, in 1967 (featuring Bob Heironimus wearing a suit Patterson purchased from my friend Phil Morris, the magician and costumer [Nickell 2011, 68–73]). As Smithsonian primate expert John Napier said regarding Patterson's creature, "I could not see the zipper" (1973, 95). With Ford's creature, one can hardly see the suit itself.

References

Holyfield, Dana. 1999. *Encounters with the Honey Island Swamp Monster*. Pearl River, LA: Honey Island Swamp Books.

———. 2007. *The Legend of the Honey Island Swamp Monster: A Documentary*. DVD including Harlan Ford's Super-8 film footage.

Holyfield-Evans, Dana. 2014. *Honey Island Swamp Monster Documentations*. Slidell, LA: Honey Island Swamp Books.

Napier, John. 1973. *Bigfoot: The Yeti and Sasquatch in Myth and Reality*. New York: E.P. Dutton.

Nickell, Joe. 2011. *Tracking the Man-Beasts: Sasquatch, Vampires, Zombies, and More*. Amherst, NY: Prometheus Books.

Part 2: The Hairy Man-Beasts

Chinese Ape-Men
In Science and Myth

The term *ape-man* is used in two major ways: scientifically, it designates any extinct primate having structural characteristics that are intermediate between man and ape (*Webster's* 1980); popularly, the term also describes any of the various legendary hairy creatures that are characterized as "human-like apes or apelike humans"—the North American Sasquatch/Bigfoot, for example (Clark and Pear 1997, 207, 467). As CFI's visiting scholar in China during October 2010 (in an exchange program with the China Research Institute for Science Popularization [CRISP]), I encountered—so to speak—an example of each of these two types of ape-man, which some believe are related. As we shall see, each has proved elusive in its own way.

Extinct Primates

One of my excursions out of Beijing was into the cave-pocked mountainous countryside at Zhoukoudian (Figure 11.1), site of a major twentieth-century paleontological find. As a boy I had been fascinated by the fossils of ape-men. These included a species discovered at this site in the 1920s, which I perused in a book given to me by my late geologist uncle, Charles Cunard, titled *Historical Geology* (Dunbar 1949). My interest was renewed during my stay in China, and I visited the Beijing Museum of Natural History with my colleague and friend, Hu Junping. Although the "Peking Man" exhibit was closed at the time, we were able to gain access to it, and I resolved to then make a pilgrimage to Zhoukoudian.

Figure 11.1. Southwest of Beijing, in the mountains around Zhoukoudian, the fossil primate Peking Man once flourished.

Subsequently, at that World Heritage Site, I climbed the steep trails to the caves where *Homo erectus pekinensis* once lived (Figure 11.2), possibly making fire and using tools of chipped flint. As his name indicates, this primitive man walked erect, although his low forehead and heavy brow ridges, as well as his protruding jaws and receding chin, show his apelike character (Figure 11.3).

Figure 11.2. Caves like this at Zhoukoudian provided shelter to the primitive Homo erectus pekinensis.

Figure 11.3. Heavy brow ridges are among the apelike features of Peking Man. (Photos by Joe Nickell)

In the museum at the site, I gazed at the famous fossils of Peking Man—actually copies; the loss of the originals represents one of the unsolved mysteries of the twentieth century. The story begins during World War II when, for safekeeping, numerous bones and teeth and five skulls were packed in wooden crates and entrusted to the U.S. Marine Corps. They were intended to be transported to the United States, but when Japan bombed Pearl Harbor on December 7, 1941, the fossil treasure disappeared. According to *National Geographic Traveler: Beijing* (Mooney 2008),

> The fate of Peking Man has been the subject of much speculation. According to one theory, the boxes went to a sea grave when the *Awa Maru* was sunk by the Americans in the Taiwan Strait. An under-water search of the site found nothing. Then, in 1966, a Japanese soldier "admitted" on his deathbed that he buried the bones under a tree in

Ritan Park . . . at the end of the war, but they were not found there either. In 2005 the Chinese government announced a new investigation. It seems that the beguiling mystery of Peking Man continues. (For more on Peking Man, see Feder 1996, 129, 149–150.)

While Peking Man stood only about five feet tall, another early resident of what is now China (as well as India and Vietnam) was comparatively huge: an extinct genus whose name means "giant ape," *Gigantopithecus*. The species *Gigantopithecus blacki* represents the largest primate of all time—a Bigfoot-sized creature, perhaps, weighing over a quarter ton. The first fossilized teeth were found in a Chinese apothecary shop in 1935; further remains—mandibles and teeth—were found at Chinese excavation sites, including cave deposits. Unfortunately, given the absence of pelvic and leg bones, the ape's size is only an estimate, and its mode of locomotion is disputed—although the dominant scientific view is that it walked on all fours (Napier 1973, 173–92; Daegling 2004, 13–16; Feder 1996, 71).

Yeti, et al.

Some cryptozoologists (those who study hidden or unknown animals) believe that fossil ape-men may not be extinct after all and indeed may be the source for reports of Bigfoot-like creatures. Such hairy man-beasts are found across Asia—if sightings, footprints, and other traces are to be believed. They include the Siberian *Mirygdy* (and its eastern, often-clothed relative, the *Chuchunaa*), the Mongolian *Alma*, the Vietnamese *Nguoi rung*, the Malaysian *Sakai*, the Nepalese *Teh-lma*, and others (Coleman and Huyghe 1999, 110–139).

Some of these, such as the *Chuchunaa*, have been supposed to be Neanderthals (*Homo sapiens neanderthalensis*, extinct humans who lived in Eurasia from about 250,000 to 45,000 years ago). This, even though the legendary creature's reported height of some six feet six inches is a foot greater than that of the average Neanderthal (Coleman and Huyghe 1999, 116). *Homo erectus*—as represented by Peking Man and the related Java Man—has also been considered a hypothetical living fossil (Krantz 1992, 186; Napier 1972, 183, 192). Still larger reported ape-men have invited comparison with *Gigantopithecus* (Heuvelmans 1972, 107; Krantz 1992, 188–93).

China's best-known man-beasts are the *Yeren* and the Yeti. The Yeti is the legendary wild man of the Sherpa tribespeople of the Himalayan Mountains, which includes Tibet, an autonomous region of southwestern China. Also known to western explorers and mountaineers as the Abominable Snowman, the Yeti reportedly ranges from the height of a normal man to eight feet tall. Covered with hair, it is also described as having a conical head and large feet. It is known largely through dubious sightings and photographs of its alleged footprints in snow, which fail to constitute credible evidence (Nickell 1995, 224–226).

Bernard Heuvelmans, known as the "father of cryptozoology" and author of the cryptozoological classic *On the Track of Unknown Animals*, suggested that Yetis might be a surviving Gigantopithecus population. Under attack from man and unable to live any longer in trees, the giant apes could have sought a safe and suitable habitat in the Himalayas, Heuvelmans hypothesized (1972, 97). Others, including anthropologist Grover Krantz (1992, 191–192), postulated that Gigantopithecus subsequently extended its range and evolved into the North American Sasquatch/Bigfoot. However, there are many arguments against either possibility, including an absence of fossil record in each case and the fact that paleontologists have concluded that Gigantopithecus became extinct about a hundred and fifty thousand years ago (Napier 1973, 178–80; Daegling 2004, 15).

The *Yeren*

As to the *Yeren*, it has existed in some version or other in the folklore of southern and central China since ancient times. Its name means "wild man," but it has also been characterized as a "manbear," "mountain monster," "monkeylike" creature, "red-haired mountain man"—even a mountain "ghost" (Coleman and Clark 1999, 260; Poirier et al. 1983, 31). Perhaps the earliest reference to such a creature, dating over 2,000 years ago, is found in the poetry of Qu Yuan, which frequently mentions the *Shangui* (or "mountain ogres"). A Tang Dynasty (CE 618–907) historian, Li Yanshow, described a group of Hubei wild men, and a Ming Dynasty (CE 1368–1644) pharmacologist, Li Shizhen, reported on several types of wild men in his voluminous *Compendium of Materia Medica*. Eighteenth-century poet Yuan Mei described an entity that was "monkeylike, but not a monkey" in Shanxi Province (Topping 1981).

Searches for the *Yeren* have been carried out since at least the 1970s, but the adventurers and filmmakers involved have failed to see the fabled creature. Sightings by others, as well as the usual doubtful footprints and hair and fecal specimens, make up the bulk of the inconclusive evidence for the existence of the alleged creature. Nevertheless, during my stay in China, the newspaper *China Daily* (Guo 2010) reported, "Search for elusive ape man continues against the odds." The article told of a planned expedition by the newly constituted Wild Man Research Association, founded in Hubei and focusing on that province's remote, mountainous area known as the

Shennongjia Nature Reserve. The report prompted a response from an ornithologist who has long studied fauna in Shennongjia. He labels the search nonsense, explaining: "That location is not consistent with that of the ape man. There's a basic standard for judging whether it exists, for example, the species grouping and area of distribution. There's no area for Wild Man's activity in Shennongjia." He concluded by pointing to the failed expeditions of the 1970s and 1980s (Guo 2010).

Descriptions of the *Yeren* are exceedingly varied. The creature is reported to range in height from as little as three feet to over nine feet. It is usually said to be covered in hair, but that varies in color from grayish-brown to brown, dark-brown to brown-red, red, and even "purple-red wavy hair," as well as white (Clark and Pear 1997, 262; Poirier et al. 1983, 31–32). Its footprints allegedly range from very small to twenty-one inches or more, and some prints suggest claws (a nonprimate feature). The creature is thought to walk upright or upon all fours, be carnivorous or vegetarian, make bamboo nests or inhabit caves, and so on (Clark and Pear 1997, 260–265; Poirier et al. 1983).

Clearly, if the *Yeren* is not entirely imaginary—functioning as a sort of folkloric bogeyman—it does not have a single, simple explanation. Sightings may simply be caused by any of such animals as bears, including albino bears, and macaque monkeys, not to mention wolves, wild goats, and numerous other wildlife. Indeed, in 1980 two supposed *Yeren* shot by a hunter turned out to be the rare and endangered golden monkey, while bears have been suggested as the explanation for certain "manbear" reports, just as albino bears (a high incidence of albinism is known in Hubei province) have been put forward for "mountain ghosts." Mange can give a mysterious appearance to an ordinary creature. For example, an "Oriental Yeti" was apparently a mangy Himalayan weasel, and a Bigfoot whose story my wife Diana and I pursued in northern Pennsylvania was most likely a black bear with mange (Nickell 2011, 61–62).

Some have suggested that the wild man is some human throwback—neither *Gigantopithecus* nor Peking Man surely but possibly some oddity like those sometimes exhibited in carnival sideshows (Nickell 2005, 150–58, 202–

208). A "monkey baby," for instance, that lived in Xhin Xhan County of Hubei Province, was simply an unfortunate individual with genetic deficiencies who "walked with a shuffling gait, had a slouched back, had a low misshapen forehead, could only make sounds with no articulate speech, and grinned constantly" (Poirier et al. 1983, 30). *Yeren* researcher Frank E. Poirier—only a normally hairy westerner who is about five feet eleven inches tall—frightened some local children who "ran away horrified at their encounter with what they screamed to others was the Wildman in their midst" (Poirier et al. 1983, 37–38).

Of course, some *Yeren* sightings and other evidence may even be due to hoaxing—the work of those seeking notoriety, enjoying pranking, or hoping to boost local tourism. In any event, until a specimen is actually captured or killed, the elusive *Yeren*—like its nearby and western counterparts, the Yeti and Sasquatch/Bigfoot, respectively—will remain in the realm of myth, not in the scientific canon like *Homo erectus* and *Gigantopithecus*.

References

Clark, Jerome, and Nancy Pear. 1997. *Strange and Unexplained Phenomena*. New York: Visible Ink.

Coleman, Loren, and Jerome Clark. 1999. *Cryptozoology: A to Z. The Encyclopedia of Loch Monsters, Sasquatch, Chupacabras, and Other Authentic Mysteries of Nature*. New York: Fireside.

Coleman, Loren, and Patrick Huyghe. 1999. *The Field Guide to Bigfoot, Yeti, and Other Mystery Primates Worldwide*. New York: Avon Books.

Daegling, David J. 2004. *Bigfoot Exposed: An Anthropologist Examines America's Enduring Legend*. New York: Rowman & Littlefield.

Dunbar, Carl O. 1949. *Historical Geology*. New York: John Wiley & Sons, 507–509.

Feder, Kenneth L. 1996. *The Past in Perspective: An Introduction to Human Prehistory*. London: Mayfield Publishing Co.

Guo Rui. 2010. Search for elusive f continues against the odds. *China Daily* (October 12).

Heuvelmans, Bernard. 1972. *On the Track of Unknown Animals*. Cambridge, Massachusetts: The MIT Press.

Krantz, Grover S. 1992. *Big Footprints: A Scientific Inquiry into the Reality of Sasquatch*. Boulder, Colorado: Johnson Books.

Mooney, Paul. 2008. *National Geographic Traveler: Beijing*. Washington, DC: National Geographic. Napier, John. 1973. *Bigfoot: The Yeti and Sasquatch in Myth and Reality*. New York: E.P. Dutton & Co.

Nickell, Joe. 1995. *Entities: Angels, Spirits, Demons, and Other Alien Beings*. Amherst, New York: Prometheus Books.

———. 2011. *Tracking the Man-Beasts*. Amherst, New York: Prometheus Books.

Poirier, Frank E., Hu Hongxing, and Chung-Min Chen. 1983. The evidence for Wildman in Hubei province, People's Republic of China. *Cryptozoology* 2 (Winter): 25–39.

Topping, Audrey. 1981. Wild men of China: Scientists stalk hairy aborigines. *Science Digest* (August); cited in Poirier et al. 1983, 28.

Webster's New Twentieth Century Dictionary of the English Language Unabridged (2nd ed.). 1980. N.p.: William Collins Publishers, s.v. "ape-man."

Bigfoot as Big Myth

The hairy man-beast known as the "Sasquatch" or "Bigfoot" is now ever present in North American culture. Supposedly a throwback to our evolutionary past, it is an "ape-man" version of us just as the little-bodied, big-headed, humanoid extraterrestrial is a futuristic one. Together they represent powerful mythologies for our shrinking planet—Bigfoot as the very symbol of the endangered species and ET as the promise that we are not alone in the universe. Here we look at at least seven mythical embodiments of the evolving Bigfoot.

1) Reporting 'Wild Men of the Woods'

In early North American accounts, the antecedent of today's Bigfoot was typically called a "wild man of the woods"—a European term from as early as the sixteenth century (Nickell 2011, 44). From 1818, when the earliest known newspaper account (in the Exeter, New Hampshire, *Watchman*) referred to an animal "resembling the Wild Man of the Woods," accounts over the next century used that term or variants, such as *wild man*, *wild child*, *wild boys*, or the like (Bord and Bord 2006, 3–24).

Typically the terminology described actual humans—including genetic oddities covered with hair and long-haired hermits and deranged people—but also the orangutan or other apes (thought perhaps escaped from traveling menageries) and real or imagined mystery woodland creatures. One creature, reported in Kansas in 1869 and referred to as a "wild man or animal," had "a stooping gait" and "very long arms with immense hands or claws"—"generally" walking "on its hind legs but sometimes on all fours" (Bord and Bord 2006, 10). It was likely a bear, since bears often stand on their hind legs and even walk when in their "alert" mode (Nickell 2013). As cryptozoologist Jeff Meldrum (2006, 204) concedes, "In behavior and appearance, no other animal is more subject to anthropomorphism than is the bear."

In the late 1830s, a "wild child" was reported swimming in an Indiana lake. In the 1860s, a Nevada creature was spotted carrying a rabbit and a club. A few others were similarly armed, including a six-foot bearded "wild man." Another "wild man" had "long matted hair and a beard," and so on. Such cases were reported well into the twentieth century (Bord and Bord 2006, 218–229). A few were allegedly captured—notably "Jacko," a hairy "half man, half beast" who stood only four feet seven inches tall. It was supposedly apprehended in 1884 by railway men and kept in an area jail (as reported in a Victoria, British Columbia, Canada, paper), but the story appears to have been a reporter's hoax (Nickell 2011, 57).

In 1924, some prospectors in Washington State had their cabin pelted with rocks by "mountain devils" (rumored to have been pranksters), and in the same year a man named Albert Ostman claimed he was kidnapped and held by a family of wild creatures (although he did not tell his tale until 1957) (Daegling 2004, 67–70).

Much earlier, in 1871, *The New York Times* had seen a trend: "As most of our readers are probably aware, there is at present roaming over the United States, and for aught we know, making occasional excursions into British America [Canada] and Mexico, a singular creature known as the 'Wild Man.'" The entity was characterized by seeming to be almost everywhere and having the "peculiar power of eluding capture" (qtd. in Arment 2006, 29).

2) Retrofitting Native American Monsters

It is common to suggest that Native Americans of the Pacific Northwest told stories and made images of hairy man-beasts similar to the modern Bigfoot. According to major proponents of this notion (Hunter with Dahinden 1993, 15), "The earliest-known 'recorded' references to the Sasquatch are found on the carved totem poles and masks of the coast Indians of British Columbia, particularly on those of the Kwakiutls. " Proponents cite such mythical creatures as the D'Sonoqua and the Buk'wus.

Actually, the Kwakiutls' D'Sonoqua were giant, man-sized cannibals who lived in houses deep in the forests. They had black bodies with hairy hands, and their eyes were deeply set. They were usually represented as females who abducted children for their tender flesh, collecting them in a basket they carried. Thus, she was a type of bogeyman—or bogeywoman. She was depicted with lips pursed so as to give her fearsome cry, "Hu! Hu!" (Alley 2003, 151–152; *The Spirit World* 1992, 47; Taylor 1994, 90–91). Similar tales were told by the Salish Indians of the Fraser Valley, British Columbia (Hunter with Dahinden 1993, 15).

As to the Buk'wus, another creature of Kwakiutl folklore, he was man-sized or smaller but hairy and having many supernatural features. For example, he was ice cold and could move in an instant to a location far away (Alley 2003, 151–153). The Buk'wus—and a very similar Pu'gwis of the Tsimshian people—were actually spirits having the facial features of a human corpse, such as stretched skin and lips curled away from the teeth (Taylor 1994, 90). Clearly neither was a Bigfoot.

Still, there were other supposed candidates. The Tlingits of southeast Alaska believed in a man-sized hairy creature called "Kushtakaas" or "land-otter man." Essentially such beings were men, but they had become mad—either by being lost or nearly drowning—and, growing hair over their bodies, went to live among others similarly afflicted. They sometimes walked on all fours in keeping with their otter-like transformation. Traditionally, they were feared for their ability to capture a person's soul, and thus are more analogous to zombies than Bigfoot (Alley 2003, 137–138).

The list goes on. However, whenever believers have attempted to equate Bigfoot with Native American folkloric entities, they have engaged in retrofitting (after-the-fact matching) and, indeed, an obvious "exercise in confirmation bias" (Loxton and Prothero 2013, 33). Nevertheless, this process remains essential to Bigfoot mythmaking. (For more on the Native American images and traditions, see Halpin 1983.)

3) Creating 'Sasquatch'

Although the name Sasquatch is often said to be Native American, it was actually coined in the 1920s by a British Columbia teacher and Indian agent, J.W. Burns. Some say Burns "Anglicized" a Native American term, but really his Native Coast Salish informants had several different names for various folkloric entities, and he wanted to invent a single term for all the alleged creatures. He harmonized some of the names—including Sokqueatl or Soss-q'tal—into "Sasquatch" (Coleman and Clark 1999, 215).

Burns himself quoted stories from elderly Indians who had encountered wild men. For example, one informant from the Chehalis Reserve saw what he took "at first sight to be a huge bear crouched upon a boulder," but when it stood up he saw it was a "man—a giant, no less than six and one-half feet in height, and covered with hair" (Burns 1929).

Another Native American, from the Skwah Reserve had seen several such wild people. He shot what he, too, first "took for a bear," coming out of a hole in a great cedar, but then he saw it actually looked like a nude "white boy." Thus provoked, a "wild woman" also came out. "Her face was almost negro black and her long straight hair fell to her waist. She was the height of a tall man, about six feet, but much broader," he said. She spoke to him in a dialect he understood, ending with, "you'll never kill another bear." So was it a bear and cub—in part, or all, a fantasy tale—or was it as Burns opines, one of "the Sasquatch people" with a white boy, stolen or found?

As Burns's new term began to catch on, it helped turn various folkloric concepts into an increasingly uniform one. It became the "Indian" name for anything that could be construed as a man-beast. The "wild man of the woods" was becoming a rather bearlike Sasquatch.

4) Discovering Big Footprints

The earliest North American record of potential man-beast footprints is from 1811, reported by trader/explorer David Thompson, who thought it likely "the track of a large old grizzled bear" (Hunter with Dahinden 1993, 17). Not only is a bear's hindfoot "remarkably human-like," but the hind- and forefoot may superimpose to look like the huge track of a bipedal creature (Napier 1973, 150–151). There were few reports of alleged Sasquatch footprints until 1930, when berry pickers near Mount St. Helens discovered huge humanlike tracks that encircled them. But more than half a century later, a retired logger named Rant Mullens confessed he had donned carved nine-by-seventeen-inch feet to make the tracks. Meanwhile, in 1951, in the Himalayas, a footprint of a Yeti, or "Abominable Snowman," was photographed by explorer Eric Shipton and widely circulated in the United States (later explained as a probable animal track, altered and enlarged by the melting snow) (Nickell 2011, 59, 68). The stage was now set for another watershed moment.

In 1958, a Sasquatch seemingly made several visits to a road construction site at Northern California's Bluff Creek. Its tracks were discovered by a bulldozer operator, Gerald "Jerry" Crew, a photo of whom— holding up a cast of a giant footprint—was spread by a wire service across the country. Consequently, the name "Bigfoot" (which first appeared in the *Humboldt Times* on October 5, 1958) began to become widespread.

The family of Bluff Creek road contractor Ray Wallace informed the press—after Wallace's death in 2002—that he had faked the 1958 tracks. They even produced pairs of carved feet that matched the Bluff Creek tracks (Daegling 2004, 29, 73; Coleman and Clark 1999, 39). However, the man-beast myth was now already entrenched, and, meanwhile, the 1958 Bluff Creek hoax had resulted in the pseudo-Indian term *Sasquatch* largely being replaced by the descriptive term *Bigfoot*.

Having previously been scarce, after 1958 reports of Bigfoot tracks began to proliferate. Tracks were reported with two to six toes and ranging in length from eleven and three quarters to twenty-one inches. Over the years the feet began to become rather standardized, usually having five toes and commonly measuring in the sixteen-to-eighteen–inch range (see Bord and Bord 2006, 215–310). And, in what seems likely to have been one-upmanship on the part of Bigfoot hoaxers, some of the tracks began to become more sophisticated.

During the 1960s, Bigfoot tracks tended to be rather rectangular in shape: the big toe was only a bit larger than the others and all five were "arranged almost in a straight line across the front of the rectangular foot pattern," according to skeptic Michael Dennett. Dennett (1996, 120, 122) noted that fake footprints subsequently improved in design so that the rectangular form was rarely seen anymore. Further sophisticated elements began to appear.

For example, more than a thousand tracks were left in 1969/1970 at Bossburg, Washington, by Bigfoot—or "Clubfoot" or "Cripplefoot" as the creature has been dubbed. They were ostensibly made by a creature with a congenitally deformed right foot. A Bigfoot-believing anthropology professor, the late Grover Krantz, asserted with hubris, "This requires an expert anatomist with a very inventive mind, more so than me, and I seriously doubt that any such person exists" (Krantz 1992, 83). However, anthropologist David J. Daegling observes that templates for Bigfoot tracks, both normal and deformed, were available in dozens of textbooks. "All a hoaxer had to do was have the wherewithal to scale them up, and he or she did not need to know one iota of anatomy to do so" (Daegling 2004, 87).

Again, in 1982, oversized footprints were discovered in Oregon with dermal ridges (those that on the hands produce fingerprints). Although the fact impressed many (Meldrum 2006, 249–259), it seems odd that previous creatures did not exhibit such features. The effect was that hoaxers were using more and more clever means to convince others that Bigfoot was real. A wildlife biologist and a professional tracker subsequently reported evidence of hoaxing, and Michael Dennett produced similar impressions (Dennett 1989).

5) Witnessing 'Bigsuit'

One series of Bigfoot tracks assumes special importance because of being found during "one of the most momentous events in the annals of Bigfoot hunting" (Bord and Bord 2006, 90). It began on October 20, 1967, when longtime Bigfoot enthusiast Roger Patterson—known as a "repeater" because of his frequent "discovery" of Bigfoot tracks—was riding horseback with friend Bob Gimlin at Bluff Creek (the area where Ray Wallace's hoaxed tracks had been made). Patterson had a 16mm movie camera and had announced his intention of filming the elusive creature.

It appeared, seemingly on cue, and Patterson filmed it briefly as it strode away with a seemingly exaggerated stride, as if, wrote one critic, "a bad actor were trying to simulate a monster's walk" (Cohen 1982, 17). (See Figure 12.1.) Patterson's creature, dubbed "Patty," had hairy, pendulous breasts—a feature so convincing, some thought, that it argued against the film being a hoax. However, Patterson had published in his book the year before a drawing of just such a female of the supposed species (Patterson 1966, 111).

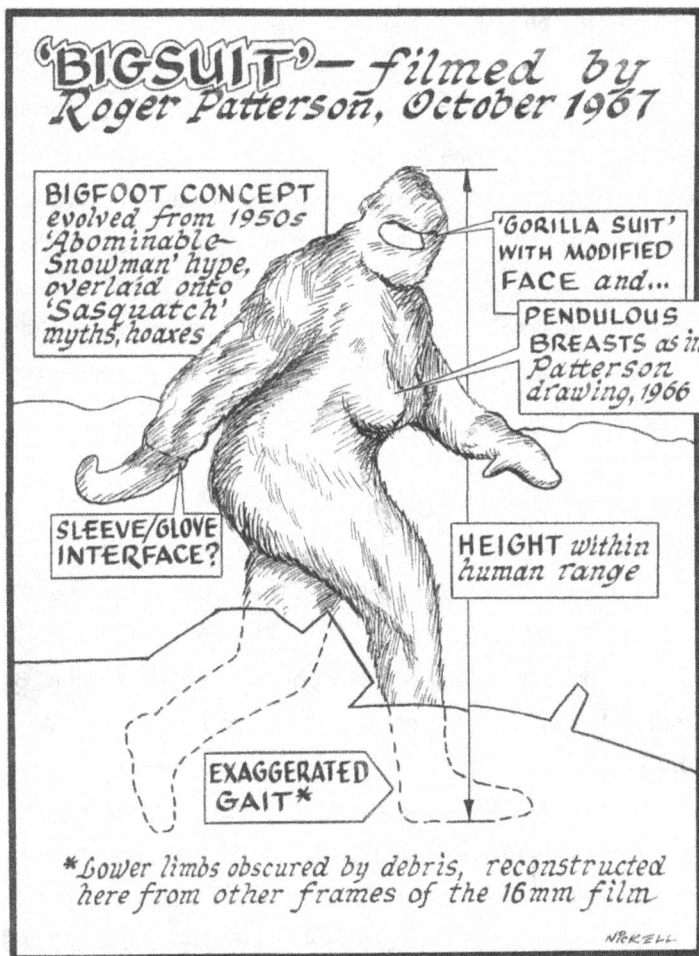

Figure 12.1. Analysis of a frame from the 1967 Patterson "Bigfoot" film shows evidence of fakery (drawing by the author). Later a Patterson acquaintance named Bob Heironimus confessed it was he who had worn the ape-man suit.

The Smithsonian Institution's John Napier analyzed the film frame by frame and concluded that the figure's walk was consistent with that of a man striding in exaggerated fashion. "The upper half of the body bears some resemblance to an ape and the lower half is typically human," wrote Napier (1973, 90–91). "It is almost impossible to conceive that such structural hybrids could exist in nature. One half of the animal must be artificial. In view of the walk, it can only be the upper half." Napier summed up, "I could not see the zipper" (1973, 91, 95).

As it happened, early in this century, a Patterson acquaintance named Bob Heironimus confessed it was he who had worn the ape-man suit, and others corroborated Heironimus's having had such a costume at the time. Also, magician-turned-costume-seller Phil Morris (whom I know and have talked with about the case on several occasions) reports that he sold a six-piece gorilla suit to Patterson, along with extra fake fur he had asked to be included. This was obviously used to transform a gorilla suit into "Bigfoot"—or rather "Bigsuit" (Long 2004; Nickell 2011, 58–72).

Nevertheless, the Patterson film became, for True Believers, long-sought-after supposed proof of Bigfoot's existence. They were supported by Grover Krantz, who believed Patterson's Bigfoot was a surviving Gigantopithecus, a "Giant Ape" of South Asia that went extinct some 150,000 years ago (Krantz 1992; 1999).

Other Bigfoot hoaxes followed, including one near Mission, British Columbia, on May 1, 1977. A few months earlier some Cashton, Wisconsin, youths admitted to a similar stunt, one dressing up as a Bigfoot-type creature with large wooden feet affixed to his shoes. Another such hoax took place in 1986 when a Pennsylvania man wore fake fur and a "wolfman" mask and alarmed nighttime drivers by appearing suddenly in their car headlights. I investigated and exposed a Bigsuit case in Western New York in 2006 (Nickell 2011, 72, 77).

A more elaborate hoax involved—as was advertised on carnival midways—a "Sasquatch Safely Frozen in Ice." It proved to be a fake made by a top Disneyland model maker. I viewed the exhibit in 1973 on the midway of the Canadian National Exhibition (where in 1969 I had worked as a magic

112

pitchman). The freezer unit was out of order, the lid up, and the ice had melted somewhat exposing part of the figure. It was dark and decidedly rubbery. This brilliant hoax was crudely imitated in 2008 by filling a Bigfoot costume with animal parts (later replaced with inorganic materials), and freezing it. It reputedly sold on eBay for a quarter of a million dollars (Nickell 2011, 87–90).

6) Connecting with Extraterrestrials

After "flying saucers" were reported in 1947 and the UFO/alien craze subsequently developed into a myth paralleling that of Bigfoot, by the 1960s the two shared the "phenomenological landscape," according to UFO historian Jerome Clark (1998, I: 469). Noting that the "hairy bipeds" seemed "like some weird marriage of apparition and animal," Clark says, some began to wonder if Bigfoot and UFOs might be related. Maybe they were "a variety of UFO occupant, possibly a lower form of life used as a sort of test animal."

During the 1970s, a number of UFOlogists—such as Coral Lorenzen, Dr. Leo Sprinkle, and Leonard Stringfield (the latter having promoted reports of crashed saucers and the secret retrieval of their humanoid occupants)— were "getting into the Bigfoot business too," say Janet and Colin Bord. They add: "There is no doubt a body of work that has Bigfoot-like creatures directly connected to UFO sightings" compiled by such UFOlogists. And authors such as Brad Steiger "were also producing paperback books full of new stories of UFOs and apemen" (Bord and Bord 2006, xi).

For example, consider some reports from a single year. Near Sykesville, Maryland, on May 29, 1973, a man claimed to see a UFO drop some object into a reservoir and then saw a luminous-eyed Bigfoot (Bord and Bord 2006, 270). Again, in October 1973 near Galveston, Indiana, one "Jeff Martin" or "Jim Mays" (the same story is told with different names) was fishing at night when he twice saw a sandy-colored Bigfoot. When it ran off, "Almost instantaneously a glowing bronze light rose from the woods and shot away into the sky."

Yet again, on October 25, 1973, near Uniontown, Pennsylvania, a man and twin boys were in a field observing a red-glowing sphere when they spotted a pair of Bigfoot creatures—first thought to have been bears—walking along a fence row (Bord and Bord 2006, 132, 274; Huyghe 1996, 70–71). Other Bigfoot/UFO links were reported over subsequent years.

There are even more far-out connections. Among alleged alien "contactees," some claim to connect with Bigfoot (Lapseritis 2014). For instance, one woman told a psychologist, "The Star People would have me read the newspaper and they would read through my eyes. They would see property that had Sasquatch on it that they would want me to buy" (qtd. in Clark 1998, I: 473). A writer for *Fate* magazine asserts that "the Sasquatch are actually extraterrestrials"—descendants of evolved "nature people" who migrated to Earth millions of years ago (Lapseritis 2014).

7) Entering Mystical Dimensions

Among the silly pretensions of "clairvoyant" Lorraine Warren (widow of Ed Warren, the notorious "demonologist" and supernatural huckster [Nickell 2012, 283–286]) is her story about telepathically communicating with Bigfoot. It happened "one spring," she says, "when we were lecturing in Tennessee and a reporter . . . told us about some hill people who kept insisting that something was threatening their children. . . . " Warren was in the fearsome bogeyman region when, standing beside a tree, she had a psychic vision of a shaggy-haired intelligent creature who had the "ability to project images telepathically into Lorraine's mind." Bigfoot told her he had injured his foot, which would keep him from the "secret cave" where his mate and children waited. Fortunately, she was able to send the creature healing images (Warren and Warren with Chase 1989, 35–43).

Warren (who exhibits several traits in common with a fantasy-prone personality [Nickell 2012, 347–348]) is not alone in believing Bigfoot has such abilities. "Let me tell you something," insists Bigfooter Ron Patillo. "These creatures are psychic. If you go in there with guns with the intent to shoot one to prove that they exist, you'll never see one. They'll pick up on you before you even get there. If you want to see them, you need someone like me to help you, who also has psychic ability" (qtd. in Burnette and Riggs 2014, 151–152).

Some "researchers" claim that Bigfoot are not merely psychic but perhaps entirely supernatural—a situation that has caused Bigfoot believers to split into supernatural and "flesh-and-blood" camps. Indeed, the supernaturalists express a variety of opinions, from believing the creatures are phantoms to considering them as "demon shapeshifters or interdimensional travelers" (Burnette and Riggs 2014, 167).

Although such ideas embarrass the flesh-and-blooders, some mystics think they can explain why Bigfoot is so "peculiarly elusive": They simply opine that the supposed creature has the power of invisibility (Burnette and Riggs 2014, 164–167)! Only time will tell what other notions will surface.

*   *   *

As the foregoing shows, during its history, the hairy man-beast has evolved through at least seven mythical embodiments: Wild Man, Indian Spirit, "Sasquatch," Bigfoot, Bigsuit, UFOlogical Entity, and Mystical Being—all perhaps summed up in one: Imaginary Creature (though based in part on the upright-standing bear).

References

Alley, J. Robert. 2003. *Raincoast Sasquatch*. Surrey, BC: Hancock House.

Arment, Chad. 2006. *The Historical Bigfoot*. Landisville, PA: Coachwhip.

Bord, Janet, and Colin Bord. 2006. *Bigfoot Casebook Updated: Sightings and Encounters from 1818 to 2004*. Enumclaw, Washington: Pine Winds Press.

Burnette, Tom, and Rob Riggs. 2014. *Bigfoot*. Woodbury, MN: Llewellyn.

Burns, J.W. 1929. Introducing B.C.'s hairy giants. *MacLean's Magazine* (April 1).

Clark, Jerome. 1998. *The UFO Encyclopedia*. In 2 vols. Detroit, MI: Omnigraphics.

Cohen, Daniel. 1982. *The Encyclopedia of Monsters*. New York: Dodd, Mead.

Coleman, Loren, and Jerome Clark. 1999. *Cryptozoology A-Z*. New York: Fireside.

Daegling, David J. 2004. *Bigfoot Exposed: An Anthropologist Examines America's Enduring Legend*. New York: AltaMira Press.

Dennett, Michael. 1989. Evidence for Bigfoot? An investigation of the Mill Creek "Sasquatch prints." *Skeptical Inquirer* 13(3) (Spring): 264–272.

———. 1996. Bigfoot. In Stein 1996, 117–125.

Halpin, Marjorie M. 1983. *Totem Poles: An Illustrated Guide*. Vancouver, B.C.: University of British Columbia Press.

Hunter, Don, with Rene Dahinden. 1993. *Sasquatch/Bigfoot: The Search for North America's Incredible Creature*. Toronto: McClelland & Stewart.

Huyghe, Patrick. 1996. *The Field Guide to Extraterrestrials*. New York: Avon Books.

Krantz, Grover. 1992. *Big Footprints*. Boulder, CO: Johnson Books.

———. 1999. *Bigfoot/Sasquatch Evidence*. Surrey, BC: Hancock House.

Lapseritis, Kewaunee. 2014. Sasquatch: A terrestrial-extraterrestrial? *Fate* 726: 8–14.

Long, Greg. 2004. *The Making of Bigfoot*. Amherst, NY: Prometheus Books.

Loxton, Daniel, and Donald R. Prothero. 2013. *Abominable Science! Origins of the Yeti, Nessie, and Other Famous Cryptids*. New York: Columbia University Press.

Meldrum, Jeff. 2006. *Sasquatch: Legend Meets Science*. New York: Tom Doherty Associates.

Napier, John. 1973. *Bigfoot: The Yeti and Sasquatch in Myth and Reality*. New York: E.P. Dutton.

Nickell, Joe. 2011. *Tracking the Man-Beasts: Sasquatch, Vampires, Zombies, and More*. Amherst, NY: Prometheus Books.

———. 2012. *The Science of Ghosts*. Amherst, NY: Prometheus Books.

———. 2013. Bigfoot lookalikes: Tracking hairy man-beasts. *Skeptical Inquirer* 37(5) (September/October): 12–15.

Patterson, Roger. 1966. *Do Abominable Snowmen of America Really Exist?* Yakima, WA: Franklin Press.

The Spirit World. 1992. The American Indians series; Alexandria, VA: Time-Life Books.

Stein, Gordon, ed. 1996. *The Encyclopedia of the Paranormal*. Amherst, NY: Prometheus Books.

Taylor, Colin F., ed. 1994. *Native American Myths and Legends*. New York: Smithmark.

Warren, Ed, and Lorraine Warren (with Robert David Chase). 1989. *Ghost Hunters*. New York: St. Martin's Paperbacks.

Sasquatch Lookalikes

Although Sasquatch—after 1958 generally called Bigfoot—is most associated with the Pacific Northwest (a region loosely ranging from northern California to Oregon, Washington, British Columbia, and southern Alaska), sightings are reported throughout the United States and Canada (Bord and Bord 2006). Many of these turn out to be hoaxes—notably Roger Patterson's filming of "Bigsuit" in 1967. (He used a gorilla suit purchased from costume-seller Phil Morris, converted it to Bigfoot by modifying the face and adding pendulous breasts, and enlisted a man named Bob Heironimus to wear the suit [Long 2004; Nickell 2011, 68–73].) Many other Bigfoot sightings are no doubt misperceptions resulting from expectation and excitement (Nickell 2011, 94–96).

But misperceptions of what? Over my years as a skeptical cryptozoologist, I have looked for real, natural lookalikes to explain various reported "monsters." For example, the round-faced, gliding, "Flatwoods Monster" of 1952 with its "terrible claws" seemed almost certainly to be a barn owl, just as "Mothman" of 1966, with its large, shining red eyes, could be identified as a barred owl (Nickell 2011, 159–66, 175–81). Again the legendary "giant eel" of Lake Crescent, Newfoundland, was probably inspired by otters swimming in a line (who are also known to be mistaken for some lake and sea monsters) (Nickell 2007; 2012a). Given these and other examples of monster lookalikes—I think of my work in this regard as that of a paranatural naturalist—we may ask: Are there animals that might be mistaken for Bigfoot?

A Candidate

As it happens, there is one especially good candidate for many sightings of Bigfoot—even for some of the non-hoaxed imprints of his big feet. The earliest record of potential Sasquatch footprints comes from an explorer named David Thompson, who while crossing the Rockies at what is today

119

Jasper, Alberta, came upon a strange track in the snow. Measuring eight by fourteen inches, it had four toes with short claw marks, a deeply impressed ball of the foot, and an indistinct heel imprint (Green 1978, 35–37; Hunter with Dahinden 1993, 16–17).

The claws do not suggest the legendary man-beast. Indeed, John Napier, a primate expert at the Smithsonian Institution and author of *Bigfoot* (1973, 74), thought the print could well have been a bear's (whose small inner toe might not have left a mark). Thompson himself thought it likely "the track of a large old grizzled bear" (quoted in Hunter with Dahinden 1993, 17).

But what about sightings? It is not uncommon for eyewitnesses to state that at first impression their Bigfoot looked like a bear, thus proving the similarity (see Figure 13.1). Yet many go on to rule out that identification, based on some aspect of appearance or behavior. However, as considerable evidence in fact shows, many Sasquatch/Bigfoot encounters may well have been of bears. Mistaken identifications could be due to poor viewing conditions, such as the creature being seen only briefly, or from a distance, in shadow or at nighttime, through foliage, or the like—especially while the observer is, naturally, excited. Non-expert observation is also problematic, as is expectancy, the tendency of people who are expecting to see a certain thing to be misled by something resembling it (Nickell 2012, 347).

Figure 13.1. Bears routinely stand on their hind legs, which – when the bear is obscured by leaves and limbs – lends itself to the illusion that a primate is being observed. (Standing bear photo copyright 2021 Erin McAllister and Zachary Allen, used with permission.)

Comparisons

A published compilation of 1,002 American and Canadian Sasquatch/ Bigfoot reports from 1818 to November 1980 is instructive (Bord and Bord 2006, 215–310). Analysis of the cases (which are presented as brief abstracts) reveals that not only general anatomy but also color variations, footprints, behavior, and geographical distribution of Sasquatch/Bigfoot are often quite similar to those of bears.

Anatomy. Bigfoot is typically described as a large, hairy man-beast. It is said to walk on two legs, to have long arms, large shoulders, and, often, no neck. Although it is frequently likened to an ape, it has been reported many times to have claws (Bord and Bord 2006, 215–310; Wright 1962).

Like Bigfoot, bears can appear as large, big-shouldered, hairy, manlike beasts. Their anatomy is consistent with bipedal standing (hence the long "arms") though much less so with walking—and, according to the Smithsonian expert John Napier (1973, 62), "At a distance a bear might be mistaken for a man when standing still. . . ." Consider this incident of a creature on the porch of a ranch house in western Washington State in 1933 (related at second hand, years later, by the daughter of the woman who observed it):

121

It was moonlight outside, and at first she thought it was a bear on the porch, but this animal was standing on its back legs and was so large it was bending over to look in the window. She said it appeared over 6 feet tall and it didn't look like a bear at all in the moonlight. She said in a few minutes it walked over [no doubt only a couple of steps] and jumped off the porch and started around the house. She went into the kitchen so she could get a good look and she said it looked just like an ape. (Lund 1969)

Ape, Bigfoot, bear? You decide, but remember, this was bear territory. And a standing black bear can be up to seven feet tall (Yosemite 2013).

During several days in April 2013 in New York State's massive Adirondack Park, where there are scattered Bigfoot encounters, I talked to hunters and others who had witnessed standing bears. One man, at whose remote home I boarded for an evening, told me of once standing face to face with a black bear: it was on its hind legs looking in the window at him!

The often-reported action of Bigfoot running on all fours is entirely consistent with a bear, as in a case of late April 1897. Near Sailor, Indiana, two farmers witnessed a man-sized beast covered with hair walking on its hind legs, but it "afterwards dropped on its hands and disappeared with rabbit-like bounds" (Bord and Bord 2006, 23, 221). No doubt the "hands" were really paws. Again, in 1970, a Manitoba man saw a seven-foot, dark Bigfoot "stand up" by the roadside at night. And in 1972, at an Iowa state park, a seven-foot brown Bigfoot was shot at and "ran away on all fours" (Bord and Bord 2006, 260, 264; see also Green 1978, 246, 178).

One Bigfoot report was inspired when, in April 1978, a Maryland farmer saw a "bear" walking upright across a field, followed by two "smaller creatures on all fours" (Bord and Bord 2006, 300). This is consistent with a mother bear in alert mode with cubs. Bears often stand on their hind legs to look and to sniff the air, and black bears usually have a litter of two, born in January or early February ("Black Bears" 2013; Whitaker 1996, 703). And so, apparently, a *stated* bear encounter was converted by enthusiasts into a sighting of "Bigfoot." Some months earlier, in the fall of 1977, two South Dakota boys (ages twelve and nine) saw only "long hairy legs" in the bushes

122

(Bord and Bord 2006, 294), and that likely bear became another "Bigfoot."
Reports of Bigfoot's gait as "peculiar" or the like (Bord and Bord 2006, 290, 291) could be consistent with the awkward gait of an upright bear.

Figure 13.2: Brown bears standing on their hind legs make a compelling natural explanation for some Bigfoot sightings. (Godi Photo/Shutterstock)

Coloration. Like descriptions of Sasquatch/Bigfoot, black bears can not only be black but also dark brown, brown, cinnamon, blond, off-white, and white (Herrero 2002, 131–34). The same is true of grizzly (brown) bears (*Ursus arctos*), which—just like a Bigfoot reported in northern California (Bord and Bord 2006, 246)—often has dark-brown, silver-tipped hair (Herrero 2002, 133; Whitaker 1996, 706).

"To confuse the novice further," states a noted authority, "there are also variations in color patterning on the coats of each species." This is due to genetic factors and to molting. With most bears, a lightening in the color of the coat occurs between molts (Herrero 2002, 133, 134).

In nighttime sightings, color may go unreported, but the animal's eyeshine is frequently described. There are numerous reports of "gleaming eyes," "large glowing eyes," "green shining eyes," "glowing amber eyes," and the like, including occasionally "red eyes" (Bord and Bord 2006, 259–300). Generally, bear eyeshine is reported as ranging "from yellow to yellowish orange, though some people report seeing red or green" ("Backpacker" 2013). The North American Bear Center mentions a black bear with mismatched eyes, due to an injured eye that "shines red rather than yellow" ("Mating" 2013).

Footprints. Bigfoot has been reported to leave tracks that had two to six toes and ranged in length up to twenty or more inches (Bord and Bord 2006, 215– 310). Of course, many large tracks—like the fourteen-inch ones of Patterson's "Bigsuit" creature—are hoaxed (Nickell 2011, 66–75; Daegling 2004, 157–87).

As to bears, Napier (1973, 150–51) observes that "The hindfoot of the bear is remarkably human-like," and that near the end of summer when worn down, the claws "may not show up at all" in tracks. Also at moderate speeds the hindfoot and forefoot prints may superimpose to "give the appearance of a single track made by a bipedal creature" (Napier 1973, 151).

Bears' five-toed hindprints range from about seven to nine inches long for the black bear to approximately ten to twelve for the grizzly (brown) bear, although some can be more than sixteen inches, and "In soft mud, tracks may be larger" (Whitaker 1996, 704, 707). As bear expert Herrero cautions: "I don't give measurements because track size varies so much depending on substratum. If a track seems very large, look at other track characteristics."

A bear's smallest toe (the innermost one, as opposed to that of humans) "may fail to register" (Whitaker 1996, 704), no doubt explaining many four-toed "Bigfoot" tracks. As well, "In mud a black bear's toe separation may not show" (Herrero 2002, 178), possibly giving rise to the illusion that—depending on just where there might be a slight separation—a "four"-toed track might appear to have been made with only two very broad toes, or even perhaps three. Rare, six-toed tracks (unlikely for either Bigfoot or bear) were found in Iowa in 1980 after a witness saw a "strange creature on all fours eating [a] carcass" (Bord and Bord 2006, 307). Except for the tracks (which were probably due to some anomaly like the overlappg of hind and fore feet), the creature is consistent with a bear.

None of the tracks mentioned in the 1,002 abstracts under study, representing reports from 1818–1980 (Bord and Bord 2006, 215–310), were reported to have dermal ridges (the friction ridges of, for instance, fingerprints). These are common to both apes and man, as well as,

124

presumably, to an ape-man. (Although in 1982, a U.S. Forest Service patrolman discovered such prints in Oregon's Blue Mountains, in the Mill Creek Watershed, noted Bigfoot skeptic Michael Dennett [1989] turned up evidence that those tracks were part of an elaborate hoax.)

Behavior. Bigfoot's reported actions are quite varied. Aside from such outlandish reports as of a Sasquatch treating an Indian for snakebite or kidnapping people, numerous acts attributed to the fabled creature again have a ready explanation: bears. For example, Bigfoot often eats berries, fruit, grubs, vegetation such as corn, fish, animal carcasses, and human rubbish. It may be seen day or night. It often visits campsites, like one raided by a "cinnamon-colored Bigfoot" in Idaho in the summer of 1968 that left tooth marks on food containers. It also peers into homes and vehicles, and sometimes shows aggression (Bord and Bord 2006, 215–310; Merrick 1933).

Similarly, bears share these and other aspects of behavior with Bigfoot. For example, bears feed on most nonpoisonous types of berries (which they eat by moving their mouths along branches). As well, they tear open rotten logs for grubs, and they feed on fruit, corn, and other vegetation, fish, live or dead mammals, and human rubbish (Herrero 2002, 183, 149–71, 47; Whitaker 1996, 708). Bears likewise are encountered both day and night (Herrero 2002, 170; Whitaker 1996, 703–709). They visit homes, vehicles, and campsites looking for food, and they sometimes show aggression (Herrero 2002, 83–87; Whitaker 1996, 703–709). These and other parallels with Bigfoot are striking.

Then there are Bigfoot's vocalizations—many of which could well be those of bears. For example, Bigfoot often growls (Bord and Bord 2006, 237, 256, 268). One "snarled and hissed" at witnesses (268), and another "chattered its teeth" (255), while others "screamed" when shot at (247, 252). Similarly, bears growl and snort, and they make loud huffing or puffing noises (Herrero 2002, 15, 16, 115). Their most common defensive display is "blowing with clacking teeth"; as well, they may bawl (from pain), moan (in fear), bellow (in combat), and make a deep-throated, pulsing noise (when seriously threatened). Cubs "readily scream in distress" (Rogers 1992, 3–4).

Distribution. The habitat of Bigfoot in the 1,002 abstracts we are

125

studying—from 1818 to November 1980—is extensive. It includes most continental American states (excepting Delaware, Rhode Island, and South Carolina) and eight of thirteen Canadian provinces. The greatest number of sightings were in Washington State (110), followed by California (104), British Columbia (90), and Oregon (77)—that is, in the Pacific Northwest, the traditional domain of Sasquatch—followed by Pennsylvania and Florida (42 each). It is reportedly seen in woods and fields, along streams, and so on (Bord and Bord 2006, 215–310; Nickell 2011, 225–29).

The distribution of black bears is strikingly similar, as shown by population maps provided by the Audubon Society (Whitaker 1996, 704) and elsewhere (Herrero 2002, 80). America's grizzly population was once quite extensive and included the western states (Herrero 2002, 4); however, grizzlies are now relegated mostly to Yellowstone Park (chiefly in northwest Wyoming) and its vicinity, and to portions of the northernmost areas of Washington, Idaho, and Montana, as well as most of British Columbia, Northwest Territories, the Yukon Territory, and Alaska (Herrero 2002, 4; Whitaker 1996, 708). Like Bigfoot, bears are also seen in woods and fields, along streams, and so on.

Assessment

Again and again come eyewitness reports of Bigfoot that sound like misreports of bears. In Washington State, for instance, in 1948, a man saw a "thin, black-haired, 6-ft Bigfoot squatting on [a] lake shore." In September 1964 a Pennsylvania man spotted "Bigfoot peering in a window of his mother's home at dusk," while a man sleeping in his car in northwest California was "woken by Bigfoot shaking it." In July 1966, a British Columbia woman saw "head and shoulders of Bigfoot above 6-ft raspberry bushes at night." In June 1976, three Floridians saw a creature "6 ft tall, with long black hair, standing in a clump of pine trees." In August 1980, two Pennsylvania men "Driving down a mountain, saw husky black hairy creature standing in road." And so on, and on (Bord and Bord 2006, 230, 241, 244, 287, 309).

126

Let it be understood that I am in no way saying that all Sasquatch/Bigfoot sightings involve bears. After all, some are surely other misidentifications or hoaxes involving people in furry suits (Nickell 2011, 72–73). As well, Venezuela's "Loy's Ape" of the 1920s was identified as a large spider monkey, and two specimens of China's legendary Yeren, shot in 1980, proved to be the endangered golden monkey (Nickell 2011, 85–87, 96).

I am merely pointing out, what should now be obvious, that many of the best non-hoax encounters can be explained as misperceptions of bears. Of creatures in North America, standing bears are the best lookalike for the bipedal, hairy man-beasts called Bigfoot. Bears also frequently behave like Bigfoot, and they are found in regions common to the legendary creature—no certain trace of which, in the fossil record or otherwise, has ever been discovered.

References

Backpacker Blogs. 2013. Ask a bear: What color are your eyes at night? Online at http://www.backpacker.com/ask_a_bear_night_eyes_shine/blogs/1944; accessed April 2, 2013.

"Black Bears." 2013. Online at http://www.catskillmountaineer.com/animals-bears.html; accessed March 25, 2013.

Bord, Janet, and Colin Bord. 2006. *Bigfoot Casebook Updated: Sightings and Encounters from 1818 to 2004*. N.p.: Pine Winds Press.

Daegling, David J. 2004. *Bigfoot Exposed*. NY: AltaMira Press.

Dennett, Michael. 1989. Evidence for Bigfoot? An investigation of the Mill Creek 'sasquatch prints.'" *Skeptical Inquirer* 13(3)(Spring): 264–72.

Green, John. 1978. *Sasquatch: The Apes Among Us*. Saanichton, BC: Hancock House.

Herrero, Stephen. 2002. *Bear Attacks*, rev. ed. Guilford, CT: The Lyons Press.

Hunter, Don, with René Dahinden. 1993. *Sasquatch/Bigfoot*. Toronto: McClelland & Stewart.

Long, Greg. 2004. *The Making of Bigfoot*. Amherst, NY: Prometheus Books.

Lund, Callie. 1969. Letter to John Green, quoted in Bord and Bord 2006, 31–33.

Mating Battle. 2013. Online at http://www.bear.org/website/bear-pages/black-bear/reproduction/14-mating-battle-combatants. html; accessed April 3, 2013.

Napier, John. 1973. *Bigfoot: The Yeti and Sasquatch in Myth and Reality*. New York: E.P. Dutton.

Nickell, Joe. 2007. Lake monster lookalikes. *Skeptical Briefs* (June): 6–7.

———. 2011. *Tracking the Man-Beasts*. Amherst, NY: Prometheus Books.

———. 2012a. *CSI Paranormal*. Amherst, NY: Inquiry Press.

———. 2012b. *The Science of Ghosts*. Amherst, NY: Prometheus Books.

Rogers, Lynn L. 1992. *Watchable Wildlife: The Black Bear*. Madison, WI: USDA Forest Service, North Central Station Distribution Center.

Whitaker, John O., Jr. 1996. *National Audubon Society Field Guide to North American Mammals*, rev. ed. New York: Alfred A. Knopf.

Wright, Bruce S. 1962. *Wildlife Sketches Near and Far*. Fredericton, NB: Brunswick Press. Quoted in Bord and Bord 2006, 35–37.

Yosemite Black Bears. 2013. Online at http://
    www.yosemitepark.com/bear-facts.aspx; accessed
    March 25, 2013.

Bigfoot Roundup: Some Regional Variants

Having long observed that many Bigfoot sightings seem consistent with bears, I have for some time been expounding on the subject—showing that, when bears stand upright on their hind legs, they become North America's foremost Bigfoot lookalikes (other than for people in Bigfoot suits). Bigfoot also usually behaves like a bear and is typically found in bear territory (Nickell 2013a).

The resemblance is sometimes made especially clear when we look at regional subtypes of Bigfoot—the Skunk Ape of Florida, for example (Figure 14.1), or a unique "bluish" creature seen in the Yukon. As it happens, the former, together with its bad odor, is consistent with the black bear (Nickell 2013b), and the latter is illuminated by the fact that black bears of a bluish color may be found in the southwest Yukon (Nickell 2014; Gloia 2011).

Here we look at several more of these regional variants of Bigfoot—such well-known creatures as Old Yellow Top, the Traverspine "Gorilla," the Fouke Monster, Big Muddy, and others—to compare their appearance and behavior with those of bears.

Figure 14.1. Dr. Gary A. Stilwell, who guided the author into the Florida Panhandle wilderness (including the Tate's Hell region and the Apalachicola National Forest), looking for the "Bigfoot Bear" (Author's photo, October 2011).

Old Yellow Top

In 1906 at a mine near Cobalt, Ontario, a group of men saw a creature that would become known as Old Yellow Top because it was described as having a light-colored mane.

Seventeen years later, in July 1923, two miners working on their claims in the Cobalt area saw "what looked like a bear picking in a blueberry patch" (Green 1978, 248–249). One stated: "It kind of stood up and growled at us. Then it ran away. It sure was like no bear that I have ever seen. Its head was kind of yellow and the rest of it was black like a bear, all covered with hair" (qtd. in Green 1978, 249).

Actually, Black Bears (*Ursus americanus*) love blueberries, are indeed completely covered in hair—which may be all or partially blond—and often stand on their hind legs to better sense something that has attracted their attention ("Black Bears" 2013; Herrero 2002, 87). (A photograph of such a blond-maned bear—although a brown bear in this instance—is shown in Herrero 2002, 133.) This ability of bears to stand upright "no doubt influenced some people's perception of them as being humanlike . . .," according to

132

Herrero (2002, 139).

A similar creature (not the same one of course) was seen again in the same area—in April 1946, by a woman with her son, and another in August 1970, by a bus driver. "At first I thought it was a big bear. But then it turned to face the headlights and I could see some light hair, almost down to its shoulders. It couldn't have been a bear," he concluded.

A passenger on the bus stated that it "looked like a bear to me at first, but it didn't walk like one. It was kind of stooped over. Maybe it was a wounded bear, I don't know" (qtd. in Green 1978, 249). Both men's first thoughts were no doubt correct, as they themselves would probably have recognized had they been more familiar with bears' stances and color phases.

Traverspine 'Gorilla'

A famous 1913 sighting of the Traverspine "Gorilla," named after a community in Labrador, occurred when a little girl saw a huge, dark-haired creature come out of the woods. "It was about seven feet tall when it stood erect, but sometimes it dropped to all fours." It left tracks in the mud, and later in the snow, "twelve inches long, narrow at the heel, and forking at the front into two broad, round-ended toes" (Merrick 1933).

Again, this is consistent with a black bear. Yes, such bears have five rather than two toes (as do most "Bigfoot," based on their alleged tracks); however, we learn that "in mud a black bear's toe separation may not show" (Herrero 2002, 178). Given the clue that the "two" Bigfoot toes were "broad," the likely explanation is that separation only appeared between the second and third toes. That would give the appearance that there were just two notably broad toes. That the heel was described as "narrow," characteristic of a bear's hind foot (Napier 1973, 61), also helps to further identify the tracks as probably a bear's. (The estimated twelve-inch length of, presumably, the hind foot is uncommonly large, but the tracks may have been overlaps of front and hind feet or the size could have been overestimated.) Besides, size can vary, that of the same foot impression being "different in the mud, in snow, or dry ground" (Herrero 2002, 175). The estimated standing height of the "gorilla" seems about right too, since a black bear can easily be seven feet tall—with or without a little girl's misperception or exaggeration.

One account tells of two creatures, one supposedly smaller than the other, yet contradictorily stating, "They sometimes stood erect on their hind legs at which time they looked like great hairy men seven feet tall" (Wright 1962). Another indication that the creatures were indeed bears came from reports that the first creature "ripped the bark off trees and rooted up huge rotten logs as though it were looking for grubs" (Merrick 1933). Indeed, the black bear, in spring "peels off tree bark to get at the inner, or cambium, layer" and "will tear apart rotting logs for grubs, beetles, crickets, and ants" (Whitaker 1996, 705).

The Fouke Monster

Probably the first sighting of what would become known as the "Fouke Monster" after Fouke, Arkansas, near where it was sighted, was in 1953. It was not seen again until 1955 when a squirrel-hunting fourteen-year-old boy fired at it with birdshot.

He described the monster as covered in reddish-brown hair or fur, standing upright at a height of some seven feet, and having very long arms. It also had a flat nose that was dark brown. The creature "stretched, sniffing the air," then started toward the boy, who shot at it. That seemed to have no effect, and the youth ran away. In 1971 hundreds of three-toed, thirteen-and-a-half-inch tracks were found in a bean field and attributed to the monster (Green 1978, 189–191). The Fouke Monster was the inspiration for the *Legend of Boggy Creek* movies (Coleman and Huyghe 1999, 56–57; Fuller 1972, 24–28). (As an example of careless research in some quarters, one source [Matthews 2008, 110] places Foukein "Kentucky.")

Be the tracks as they may, the boy's description is a pretty good fit for an *Ursus americanus* (black bear) of cinnamon color. States one bear expert (Herrero 2002, 131–132): "An individual [black] bear's coat color may range from blond, cinnamon, or light brown to dark chocolate brown or jet black."

(See also Van Wormer 1966, 21.) Significantly, the Fouke Monster stood and sniffed the air; that is common behavior for a bear "trying to sense something" (Herrero 2002, 139), as the creature obviously was attempting in this instance.

Momo

In 1972, "Momo," short for "Missouri Monster," appeared near Louisiana, Missouri. A huge creature covered in black fur, it stood upright on two legs. Reports of such monsters date back to the 1940s, and a year prior to "the Momo scare" two women had encountered a hairy ape-man on River Road near the town (Green 1978, 194–195; Coleman and Huyghe 1999, 50–51).

Then, on July 11, 1972, an incident occurred that received national attention, with many eastern U.S. newspapers sending reporters to cover the story. About 3:30 PM on that sunny Tuesday, a fifteen-year-old girl heard her younger brother scream. Looking out a window, she saw a blood-flecked monster holding a dead dog under one arm; then it "waddled" off (Coleman and Huyghe, 1999, 50; Green 1978, 195). According to John Green (1978, 195), "after the fuss started several other people claimed to have seen something similar, generating even more excitement, and a lot of people spent time monster hunting, but nothing came of it."

I have compiled this composite description of the creature: It stood about six or seven feet tall, was neckless, and was completely covered in long black hair—even its face, according to one source. It appeared to be bipedal yet "waddled" or walked awkwardly. It had a foul smell and was, at least in part, carnivorous, capable of killing and carrying away a dog (Green 1978, 195; Coleman and Huyghe 1999, 50–51).

I submit that this is a convincing description of a black bear standing upright with its waddling gait a corroborative detail. Of course we should read "arm" as "front leg." (See Nickell 2013a.) As to the hair-covered face, a feature not reported by all witnesses, it may be that the creature was actually seen from the back. In this light, an artist's conception of Momo, reproduced in Coleman and Huyghe's *The Field Guide to Bigfoot, Yeti, and Other Mystery*

Primates Worldwide (1999, 51), strongly resembles a black bear viewed from behind.

Big Muddy

In the vicinity of the Big Muddy River near Murphysboro, Illinois, came reports of a seven-foot Bigfoot described as "dirty white" or white "with muddy body hair," or even as a "big white ghost"—from three sightings in two nights, June 25 and 26, 1973 (Bord and Bord 2006, 270). Two of the witnesses, teenagers, thought it had been covered in mud or slime from the river. Later that summer it was seen "three or four" additional times (Green 1978, 204).

"The Big Muddy Monster"—as it was now known—was seen again the following year, July 9, 1974, and again in July 1975, both times in the vicinity of Murphysboro (Green 1978, 204; Bord and Bord 2006, 270, 277, 281). On its way to legendary status, "Big Muddy" has also been styled the "Murphysboro Mud Monster" in that learned tome, *Monster Spotter's Guide to North America* (Francis 2007, 107), which tells us that it is "Seven to eight feet tall, weighing over two hundred pounds," that it is "omnivorous," and "may be dangerous if cornered or startled."

I do not doubt it. Big Muddy sounds for all the world like a tall black bear—one not black in color however. Black bears can be off-white and even white—as shown in Whitaker (1996, 703, color plate 337)—and albino ones are also known (Herrero 2002, 132).

Sister Lakes Monster

This Michigan creature eventually attracted others—including some faux monsters. In May 1964, something that seemed to lurk in the swamps nearby caused frightened fruit pickers to abandon their fields near Sister Lakes, a rural community in the state. For two years there had been reports of a mystery creature, but now they had escalated until they dominated the headlines in local newspapers. The entity was described as very tall, covered in black fur, and having eyes that "glowed with reflected light" (Bord and Bord 2006, 77). That is, the creature merely exhibited animal eyeshine, reflecting light from such sources as car headlights.

Indeed, two fruit pickers, brothers from Georgia, saw the creature in their headlights on June 9, standing upright at an estimated nine feet tall. It appeared to be a cross, they said, between a gorilla and—a bear. What this characterization probably meant was that it looked like a bear if (as they perhaps did not know bears did) one stood on its hind legs. Over the next couple of days, what is now called the Sister Lakes Monster had presumably returned to its lair (Bord and Bord 2006, 77–78; Brandon 1978, 115; Coleman and Huyghe 1999, 50).

Nevertheless, would-be monster hunters and sightseers flocked to the area, where cafes served Monster Burgers, a theater played horror fare, and the radio station interrupted its Monster Music with monster updates. One enterprising storekeeper marketed a monster-hunting kit, complete with a light, net, baseball bat, and—just in case—a wooden stake and mallet. All in good fun, but when the teenagers began to don old fur coats and imitate the monster, the sheriff stated, "I had to order hunters away because it's getting mighty dangerous—three thousand strangers prowling about at night with guns..." (Bord and Bord 2006, 78).

It would hardly have been safe for people or bears. Allowing for a little understandable exaggeration (monsters tend to loom larger, rather than smaller, in frightened witnesses' eyes), I think the Sister Lakes Monster was likely to have been, once again, a black bear.

Dwayyo

In Maryland, near wooded Gambrill State Park in 1965, one John Becker claimed that he went outside his house to investigate a strange noise, and on his way back, he said, he saw something approaching. "It was as big as a bear, had long black hair, a bushy tail, and growled like a wolf or a dog in anger." Moving toward him, it stood upright and began to attack, but Becker fought it until—with his wife and children looking on in horror—the creature ran away. He subsequently claimed to have filed a report with the Maryland State Police ("Dwayyo" 2013).

Now, this account of a four-footed creature "as big as a bear" sounds just like a bear (save perhaps for the "bushy tail"—a black bear's being only about six inches long), but in what is described almost like hand-to-paw combat, surely one would recognize a bear. Actually, this story was almost certainly a hoax. A state police spokesman denied they received any such report, and "Becker" proved to be a pseudonym.

An opportunistic reporter for the *Frederick News Post*, George May, who cranked out several articles on the creature, is implicated: Someone had used the Becker name to apply for a "Dwayyo License" from the county treasurer's office and the return address was in care of George May at the *Frederick News Post*! ("Dwayyo" 2013; Chorvinsky and Opsasnick 1988; "Big hairy monsters" 1973) Whatever the truth, some earlier and subsequent sightings of a creature in the area could well be those of bears.

Summary and Conclusions

As these several examples show, not only does Bigfoot most resemble an upright-standing bear generally, but certain well-known regional subtypes—both past and present—seem to tally with black bears: Florida's Skunk Apes (smelly black bears); a "bluish" Bigfoot in southwestern Yukon (a blue-gray or "glacier" variety black bear); the blueberry-picking Old Yellow Top of Cobalt, Ontario (a blond-maned black bear); Labrador's bipedal/quadrapedal traverspine "Gorilla" (a typical large black bear); Arkansas's reddish-furred Fouke Monster (a familiar cinnamon-colored black bear); Momo, the Missouri Monster, with "waddling" walk (an upright-walking black bear); Illinois's "dirty white" creature known as Big Muddy (a white, possibly rare albino black bear); the nine-foot Sister Lakes, Michigan, monster with "glowing" eyes (the exaggeration of a black bear with normal eyeshine); and the Dwayyo of Maryland lore (a hoax with occasional bear sightings).

Many other examples could be given, but these should be sufficient to

138

show that indeed, not only do bears often double as Bigfoot but some specific subtypes are mimicked by particular bear behavior, varieties of coloration, or other traits.

References

"Big hairy monsters." Maryland. 1973. *INFO Journal* III.3 (Summer): 27–28.

"Black Bears." 2013. Online at http://www.catskillmountaineer.com/animals-bears.html; accessed March 25, 2013.

Bord, Janet, and Colon Bord. 2006. *Bigfoot Case Book Updated*. N.p.: Pine Winds Press.

Brandon, Jim. 1978. *Weird America*. New York: E.P. Dutton.

Chorvinsky, Mark, and Mark Opsasnick. 1988. Notes on the Dwayyo. *Strange Magazine* 2: 28–29.

Coleman, Loren, and Patrick Huyghe. 1999. *The Field Guide to Bigfoot, Yeti, and Other Mystery Primates Worldwide*. New York: Avon Books.

"Dwayyo." 2013. Online at http://en.wikipedia.org/wiki/Dwayyo; accessed December 18, 2013.

Francis, Scott. 2007. *Monster Spotter's Guide to North America*. Cincinnati, Ohio: HOW Books.

Fuller, Curtis. 1972. I see by the papers. *Fate* March: 7–35.

Gloia, Carol. 2011. *Facts about the American Black Bear*. Online at http://www.critters360.com/index.php/facts-about-the-american-black-bear-4012/; accessed December 18, 2013.

Green, John. 1978. *Sasquatch: The Apes Among Us*. Saanichton, B.C.: Hancock House.

Herrero, Stephen. 2002. *Bear Attacks*, rev. ed. Guilford, CT: The Lyons Press.

Matthews, Rupert. 2008. *Sasquatch: True-Life Encounters with Legendary Ape Men*. Edison, NJ: Chartwell Books.

Merrick, Elliot. 1933. *True North*. New York: Charles Scribner's Sons. Cited in Green 1978, 252–254.

Napier, John. 1973. *Bigfoot: The Yeti and Sasquatch in Myth and Reality*. New York: E.P. Dutton.

Nickell, Joe. 2013a. Bigfoot lookalikes: Tracking hairy man-beasts. *Skeptical Inquirer* 37(4) (September/October): 12–15.

———. 2013b. Tracking Florida's Skunk Ape. *Skeptical Briefs* 23(3) (Fall): 8–10.

———. 2014. The Yukon's Bigfoot Bears. *Skeptical Briefs* 24(2) (Summer): 8–9.

Van Wormer, Joe. 1966. *The World of the Black Bear*. Philadelphia: J.B. Lippincott.

Whitaker, John O. Jr. 1996. *National Audubon Field Guide to North American Mammals*, rev. ed. New York: Alfred A. Knopf.

Wright, Bruce. 1962. *Wildlife Sketches Near and Far.* Fredericton, NB: University of New Brunswick Press.

Tracking Florida's Skunk Ape

Combining myths of the American Sasquatch—better known since 1958 as "Bigfoot"—and various swamp monsters, Florida's "Skunk Ape" is reportedly a large, shaggy, man-beast that haunts, especially, Florida's wilderness areas (Coleman and Huyghe 1999, 56–57). On a trip to the state's Panhandle region in October 2011, I was able to begin to look into the various legends and sightings—first, with a day's excursion into the remote Tate's Hell wilderness area (Figure 15.1) and part of a night in the Apalachicola National Forest with Dr. Gary A. Stilwell as guide, and, second, research trips to the state's Wildlife Commission offices and State Library and Archives of Florida in Tallahassee.

Figure 15.1. The author looking for Skunk Apes in Florida's Tate's Hell region. (Author's photo by Dr. Gary A. Stilwell)

I have since conducted much additional research on the fabled creature, which is essentially only a regional variant of the north American Bigfoot itself—see my "Bigfoot lookalikes" (Nickell 2013). (After the Pacific northwest, Florida and Pennsylvania are the most Bigfoot-reported regions of north America—at least through 1980 [Nickell 2011, 225].) In addition to

Skunk Ape, it has been called Stink Ape, Skunk Man, Skunk Monkey, Swamp Man, the Swamp Monster, and, among many others, the Bardin Booger. (the latter beast— reported in the region around the logging community of Bardin— is a sub-variant, itself having such names as Wooly Booger, Bardin Goomer, and several others, including even the Boogie Man, a name that reveals something of its status as a folk monster [Jenkins 2010, 80, 102].) Here is some of what I discovered about the Skunk Ape.

Skunk Ape Portrait

I studied a wealth of hairy man-beast encounters, selecting—from a pro-Bigfoot data base of 1,002 reports (1818–1980 [Bord and Bord 2006, 213–310)—all forty-two entries for Florida, to which I added thirty-five more from another such source (1818–2008 [Jenkins 2010, 77–128]) for a total of seventy-seven case studies. I then extracted data to determine the averages for the following characteristics of the Skunk Ape.

Physical description. The Florida Skunk Ape has generally black or "dark" long hair or fur—one report described it as seemingly "covered in fur, as if wearing a fur coat" (Jenkins 2010, 114). It may also be brown, or—in one 1848 instance—white. It has a large, round head with big, shining eyes, no appreciable neck, and broad, rounded shoulders. When standing upright, it has "long dangling arms," in one case being observed "swinging its arms as dogs yapped at it" (Bord and Bord 2006, 244).

However, it is seen in various positions: one creature was "close to the ground, as if kneeling," while another "stood up in a half crouch," then took a "huge stance with hunched shoulders"; still another was "a huge shape" that "stood up," while often the creatures were first seen standing, watching people. Estimates of its height vary greatly, from as short as four feet to as tall as ten, but the average is 7.45 feet (slightly smaller than the overall North American Bigfoot average of 7.57, determined from the 1,002 cases cited previously). Limited estimates of its weight yield an average of 508.3 pounds.[2] Its gait is sometimes said to be unusual—for instance, "exaggerated." One

144

witness said the creature "wobbled" as it walked (Jenkins 2010, 96, 105).

Odor. The Skunk Ape is supposedly distinguished as "smelly," occasionally likened to its namesake, but more often it is characterized descriptively as having a "rancid, putrid odor," like "that of rotten food and dead animal" (Bord and Bord 2006, 245; Jenkins 2010, 898), or having "the usual scent of cabbage and rotten eggs" (Jenkins 2010, 99). In fact, however, similar Bigfoot creatures across north America are also commonly described as "smelly," "strong-smelling," having a "strong animal smell," "nauseating odor," or a smell as of a "sewer" or "rotten eggs," and the like (Bord and Bord 2006, 23, 234, 247, 249, 270, 272).

Behavior. The Skunk Ape's behavior is typically similar to that of Bigfoot everywhere. It is frequently seen standing among trees, crossing a road (and occasionally being hit by a car), rummaging in garbage, drinking water or catching fish from a lake or stream, visiting campsites, standing to peer into windows, and so on. It typically vocalizes by growling, grunting, grumbling, or producing "stressed breathing" and, at least once, "clicking sounds," among others (although at times there is no sighting and so no certainty that the sound was that of a Skunk Ape) (Jenkins 2010, 111, 117, 123).

Habitat. Skunk Apes are encountered generally in remote areas, notably forests and swamps, including the everglades, as well as other national and state parks. They are attracted to human habitations—campsites, cabins and other outlying homes, and garbage dumps—in search of food (Jenkins 2010, 77–128).

Sign. Any evidence that a certain type of animal has been in a given area is called its sign. This can include tracks, indications of feeding (such as food remnants), scat (fecal matter), and the like. In the seventy-seven cases studied, the Skunk Ape's signs include large tracks, typically up to 17.5 inches and with five toes (Bord and Bord 2006, 257, 262). Other on-site indicators were broken branches, a puddle of apparent urine, and uprooted plants (Jenkins 2010, 88, 95, 101).

Suspects

As it happens, there is a known animal that actually has the foregoing characteristics: the black bear (*Ursus americanus*). It is typically covered with shiny black fur and has a tan or grizzled snout. Black bears can also be other colors, including cinnamon and even white (Herrero 2002, 131–32). A large one can stand seven feet tall (Yosemite 2013), weighing in the range of 203–587 pounds (Whitaker 1996, 703). When it stands, its "arms" dangle. it has a big head, large shining eyes, "no neck" (as is said of the Skunk Ape), and rounded shoulders.

Bears can be malodorous, and some people claim they can smell them when they are nearby (Herrero 2002, 115). Since bears often scavenge on dead animals and rummage in garbage bins and open dumps (Herrero 2002, 43, 156; Whitaker 1996, 706), they might be expected sometimes to be "smelly."Bears stand on their hind legs for various reasons, such as when necessary to peer in a window or when trying to sense something, sniffing the air. They can walk in ungainly fashion this way. States one expert, "no doubt the ability of bears to stand on two feet has influenced some people's perception of them as being humanlike . . ." (Herrero 2002, 139). Indeed, the bear's hind footprint is "remarkably human-like," especially when, in late summer, the claws are worn and "may not show up at all" in its tracks. At moderate speeds the hind and fore feet may superimpose to "give the appearance of a single track made by a bipedal creature" (Napier 1973, 150–51).

Bears behave like Bigfoot often does. They stand and watch people, visit their camps and homes, wade in streams seeking fish, climb trees for protection, and so on. They vocalize with growls, snorts, and loud huffing noises; common defensive display is "blowing with clacking teeth" (Whitaker 1996, 703–706; Herrero 2002, 15, 16, 115; Rogers 1992, 3–4).

Black bear habitat is similar to that of Bigfoot, since it consists of "primarily forests and swamps" (Whitaker 1996, 704). The big mammals once occupied all of Florida's mainland, as well as some coastal islands and the

146

larger Keys, but settlement reduced their range to scattered core areas now designated as primary range (containing core bear population) and secondary range (where bear movement is also significant although the range is less optimal) ("Black Bears" 2013). In addition to tracks, scat, and other signs, bears leave feeding signs that include broken vegetation (mangled berry patches, broken fruit-true branches, uprooted plants) and remnants of carrion and large prey.

Some Brief Case Studies

Here are a few reports of Florida Skunk Ape encounters that could be explained as misidentifications of bears:

• In 1957, in the everglades in late afternoon, a wild-boar hunter encountered "What looked like a bear squatting," but then "the thing slowly stood up to a staggering height of about eight feet." As he backed away out of the dark thicket, he glimpsed sunlight on the eyes yielding "a yellow-orange glow like the eyes of a wild animal," and the hunter ran to his truck (Jenkins 2010, 89–90.). Apparently the only thing that made him think the bearlike creature was not a bear was his mistaken belief that bears do not stand upright.

• In 1960, in a sparsely populated area near Hollywood (near the outskirts of the everglades), an "adolescent skunk ape" walked out of a drainage ditch after midnight, then stood in the center of the road. From fifty yards away, the driver of a car saw that the creature was no more than five feet tall, had long arms and a round head. It was "covered in dark fur and had no observable facial features" (Jenkins 2010, 91–92).

• In 1966, near the Andote River, a man reported seeing Bigfoot "standing in trees" and having a "rancid, putrid odor" (Bord and Bord 2006, 245).

• In 1969, near Davie, Florida, a man encountered a "smelly, growling Bigfoot" in an abandoned guava orchard, and another man saw a "huge black Bigfoot treed by dogs" in an orange grove; "it swung away through the trees," then dived into a canal (Bord and Bord 2006, 256). Black bears feed on

147

various fruit and even climb trees for food, with broken fruit tree branches being among the common signs of black-bear activity (Whitaker 1996, 703, 705). I suspect the phrase *swung away through the trees* in the account crept in because of the notion that Skunk Apes are apelike; I suggest the man misperceived how the bear made its mad scramble through the branches to the canal.

• In 1971, at Crystal River, four men saw four manlike animals on an embankment outside "a massive forest." They were picking at some plants (later found "pulled away from the earth"). The creatures were furry "from head to toe" and had "long arms and large heads that were not proportionate to their bodies" (Jenkins 2010, 100). The description is quite similar to bears, among whose feeding signs is "ground pawed up for roots" (Whitaker 1996, 703).

Conclusions

Of course not all Skunk Ape reports represent sightings of bears. Some are the product of folklore (as Jenkins [2010, 79–81] readily admits), or the misidentification of other wildlife (especially those "encounters" consisting of nothing more than sounds or eye-shine), and many could well be outright hoaxes. In fact, Bigsuit-style pranks were common regarding north-central Florida's Skunk Ape known as the Bardin Booger (Daegling 2004, 237–45).

However, considerable evidence suggests that bears, which are known to exist, can be mistaken for the Skunk Ape as well as Bigfoot in general, the existence of which lacks proof. We must recall the principle of Occam's razor (named for fourteenth-century philosopher William of Ockham), which holds that the simplest tenable explanation—the one with the fewest assumptions—is most likely to be correct.

Notes

1. Tate's Hell State Park is said to be "One of the prime habitats for the swamp-dwelling Sasquatch" (Hinson 2010).

2. For two of the cases in Bord and Bord (2006, 246, 262), I found weight data from another source ("Skunk Ape" 2013), thus making a total of six estimates of weight for all of my seventy-seven cases.

References

Black Bears distribution Map. 2013. Available at
http://myfwc.com/conservation/you-conserve/
wildlife/black-bears/distribution-map/; accessed April
22, 2013.

Bord, Janet, and Colin Bord. 2006. *Bigfoot Casebook Updated:
Sightings and Encounters from 1818 to 2004*. N.p.: Pine
Winds Press.

Coleman, Loren, and Patrick Huyghe. 1999. *The Field Guide to
Bigfoot, Yeti, and Other Mystery Primates Worldwide*.
New York: Avon Books.

Daegling, David J. 2004. *Bigfoot Exposed*. NY: AltaMira Press.

Herrero, Stephen. 2002. *Bear Attacks*, rev. ed. Guilford, Ct: the
Lyons Press.

Hinson, Mark. 2010. Florida is a haven for vampires, Skunk
Apes and Pig Men. *Tallahassee Democrat* (October 31).

Jenkins, Greg. 2010. *Chronicles of the Strange and Uncanny in
Florida*. Sarasota, Fl: Pineapple Press.

Napier, John. 1973. *Bigfoot*. New York: E.P. Dutton.

Nickell, Joe. 2011. *Tracking the Man-Beasts: Sasquatch,
Vampires, Zombies, and More*. Amherst, NY:
Prometheus Books.

———. 2013. Bigfoot lookalikes. *Skeptical Inquirer* 37(5) (September/October): 12–15.

Rogers, Lynn l. 1992. *Watchable Wildlife: The Black Bear*. Madison, WI: USDA Forest Service, North Central Station Distribution Center.

Skunk Ape. 2013. Available at http://www.weirdus.com/states/florida/bizarre-beasts/skunk-ape/; accessed April 22, 2013.

Whitaker, John O., Jr. 1996. *National Audubon Society Field Guide to North American Mammals*, rev. ed. New York: Alfred A. Knopf.

Yosemite Black Bears. 2013. Available at http://www.yosemitepark.com/bear-facts.aspx; accessed March 25, 2013.

Man-Beasts of the Yukon

Canada's Yukon Territory is a wild, rugged land, its summers having a "midnight sun," and its winters a day-long dark. Bordered on the west by Alaska, the east by the Northwest Territories, the south by British Columbia, and the north by the Beaufort Sea, the Yukon became famous for the Klondike gold rush of 1897–1898. In addition to gold, its treasures include the breathtaking northern lights and rich flora and fauna. The latter's mammals include the caribou (the same species as reindeer), moose, mountain goat, Alaskan and timber wolf, red fox, mink, otter, and many others, including the black and grizzly bear. In modern times, some say, it is also home to the legendary Sasquatch, usually known since 1958 as Bigfoot.

Bigfoot Country?

I spent most of two adventure-filled years, 1975–1976, in the Yukon, living in frontier Dawson City and working as a casino dealer, museologist, riverboat manager, and newspaper stringer, among many other activities (Nickell 2008). I have reported elsewhere on my investigations of Yukon gold dowsers (Nickell 1988, 89–102) and Dawson's "haunted" Palace Grand Theatre (Nickell 2012, 167–170).

I had not yet begun my search for what I now call the Bigfoot Bear—referring to an upright-standing bear's propensity to be mistaken for Bigfoot in general anatomy and coloration as well as behavior and geographic distribution (Nickell 2013). However, both my work and leisure put me in contact with many Yukon outdoorsmen—like riverboat captain Dick Stevenson, numerous salmon fishermen and gold miners, dog-sled-traveling trappers like Ed Wolfe and "Skipper" Mendelsohn, and many, many more, including old Joe Henry, a nationally famous Native American snowshoe maker and my favorite wintertime bar companion. I never heard mention of Bigfoot from any of these people, but they were all familiar with bears. I was often out in bear country myself, hunting, prospecting, and the like—most often with Captain Stevenson—and I once helped him as he bravely caught (by manually operating the game warden's defective mobile bear-trapping cage) a large, nuisance brown bear (Nickell 1976).

Despite a lack of convincing evidence for Bigfoot, belief persists, and Bigfoot buffs are active almost everywhere, including the sparsely populated Yukon. Indian legends are often trotted out, like the Kushtaka or "Land Otter Man" of the Tlingits of the Pacific Northwest. Kushtaka, it is said, "moves like the wind and disappears at will only to reappear again elsewhere, all the time keeping its hand before its face and peering out at times through its fingers" And "Whenever Kushtaka catches and breathes on its captive, he loses all sense of reality until the Kushtaka leaves" (Coleman 2011). Clearly Kushtaka is a kind of supernatural bogeyman of the Tlingits—not the supposedly real object of Sasquatch hunters' quests. Yukon Sasquatcher Red Grossinger, a retired Canadian Army officer, admits he has never seen a Sasquatch/Bigfoot, but he assumes an unidentified smell he experienced in 2003 may have been from one. He says his Canadian Sasquatch Research Organization (CSRO) would like to "prove its existence" (Patrick 2009), a kind of cart-before-the-horse motivation that seems a recipe for bias. In this light, let us look at some published reports of creatures that are supposed to be Bigfoot but may well be familiar creatures instead.

'Black Giants?'

As one source notes of the interest in Bigfoot:

So intense is this fascination that some Bigfoot enthusiasts seem to have labeled just about every mythological creature ever known in the western hemisphere as another name for Sasquatch. There are amusing collections of "Native American names for Bigfoot" online that include the names of giants, dwarves, ghosts, gods, underwater monsters, four-legged predators, an enormous bird, and a disembodied flying head. ("Native American" 2013)

Dolores Cline Brown, in her book *Yukon Trophy Trails* (1971, 153), along with another writer (Kristian 2013), tells of one such Yukon "Indian" legend (no further identification given) about one bogeyman known as the Bushman or Black Giant. Authentic native lore or not, this man-beast sounds curiously familiar, resembling a bear standing upright, a common posture with a "human appearance" (Van Wormer, 1966, 30). The creatures were very large and covered in black hair like the typical *Ursus americanus*, the black bear, or even like the much larger *Ursus arctos*, the grizzly (brown) bear, which can on occasion be black (Herrero 2002, 37, 133). They were said to live alone in caves or recesses during winter—like bears in their dens—and to have enormous feet (Brown 1971, 153; Kristian 2013). Bears, in fact, leave very "humanlike" hindprints—up to sixteen or more inches for a grizzly—that can appear larger in soft ground or when, at moderate gaits, hind- and forefoot prints may superimpose to appear as a single track (Nickell 2013).

A black-colored bear may well explain reports (relatively modern ones) of the Black Giant looking through cabin windows—the common behavior of bears who stand upright and peer into dwellings looking for food. It may also explain a reported instance of a presumed Black Giant heavily rattling a Yukon cabin door in the middle of the night, attempting to gain entry, which was thwarted by occupants shoving furniture against the door (Brown 1971, 153).

Reportedly, the fearsome Black Giants also occasionally "ate Indians" (Brown 1971, 153). True accounts of grizzlies and black bears eating people are gruesome indeed. One woman, carried off from camp by a grizzly, was heard to cry, "He's got my arm off," and "Oh God, I'm dead," and then was heard no more. Her body was subsequently found, partially devoured. Later, park rangers hunted down and killed the bear, actually an old female, and an autopsy revealed human hair still in her stomach (Herrero 2002, 49–50).

But why would Native Americans not recognize the Black Giant as a bear? Perhaps the putative legends got started (and we do not know how old they are) with the appearance of a rare black grizzly, standing upright (as they do in alert mode), and resembling for all the world a man-beast—indeed a black giant. To a people who believed in many imaginative beings and who were inveterate storytellers, such accounts are not at all surprising—especially if they are of modern vintage and influenced by the Bigfoot myth.

A Bearlike Creature

It pays to backtrack sources. One compendium (Bord and Bord 2006, 231) related a case from the Yukon Territory in the 1940s, in which a witness shot his .30-06 rifle at a ten-foot Bigfoot (no further description given) that reportedly left tracks eighteen to twenty-two inches long. The original source, however, had been a letter from the man, who actually admitted he was "not sure it was not a bear" (Green 1973, 17).

We need be no less skeptical than the witness himself, and the solid evidence for the existence of bears trumps that for Bigfoot, which is zero. As to the tracks, we have only the man's memory about them—a memory so lacking that he could not even remember the exact year of the event "in the 1940s." Also, recall our earlier discussion about bear tracks.

'Bluish' Bigfoot?

On October 4, 1975, a man named Ben Able reported a strange encounter near Jake's Corner, Yukon. He passed a bipedal figure on the road at night but, when he backed up to offer a ride, the figure moved away from the road. Covered with fur, it was about five and a half feet tall—generally the appearance and size of a small black bear (a standing one can be up to seven feet tall) (Bord and Bord 2006, 283; Green 1978, 242).

Curiously, however, its fur was "bluish" and it had a gray face—odd coloring for Bigfoot. However, it happens that rare "blue" or "bluish-tinged" or "blue-gray" black bears, known as "glacier" bears, are found in the Southwest Yukon, as well as nearby coastal areas of Alaska and British Columbia (Gloia 2011; Whitaker 1996, 703; Herrero 2002, 132; Van Wormer 1966, 20). This is rather unique evidence in favor of identifying the mystery creature as a regional black bear.

Encounters at Teslin

Another somewhat similar case occurred in June 2004 near Teslin (a village just north of the border with British Columbia). Two men driving a truck on the Alaska Highway after one o'clock in the morning passed a figure standing by the road. Going back to see if it was someone who needed a lift, they saw a hunched-over creature approximately seven-feet-tall, covered with dark hair. In their headlights, however, they thought they saw "flesh tones" beneath the hair. Driving away, they looked back and observed it cross the road "in two or three steps," according to a source, apparently at third hand or worse (Bord and Bord 2006, 209).

Crossing a highway in just three steps from a standing position seems remarkable even for a seven-foot man-beast. Given that the observation was made under poor conditions (behind them, without benefit of headlights, and having to use mirrors or turn awkwardly), I suspect the witnesses were mistaken. A bear seems the most likely culprit.

Some additional sightings were reported in 2005 in Teslin, culminating in the discovery of some "sasquatch hair." This was sent to a conservation office in Whitehorse, the Yukon capital ("Sasquatch" 2005). The results were good news and bad news for Bigfooters: The hair was not from a bear! Alas, it was also not from a Bigfoot—no authentic trace of which has ever been found—but from a "buffalo," that is, an American bison (Kirk 2006).

Conclusions

As the foregoing cases show, much of the evidence for Bigfoot depends on selective reporting of eyewitnesses' descriptions, the weakest kind of evidence. They may easily be mistaken due to poor viewing conditions, excitement, and even what cryptozoologist Rupert Gould (1976, 112–113) termed "expectant attention." That is the tendency to see what one expects to see—itself due, in the case of Bigfoot, either to wishful thinking or to what I call "Bigfoot programming." This refers to the fact that people are often

156

assailed with images of Bigfoot—on TV shows, for instance, far more than they are with, say, those of bears.

References

Bord, Janet, and Colin Bord. 2006. *Bigfoot Casebook Updated*. N.p.: Pine Winds Press.

Brown, Dolores Cline. 1971. *Yukon Trophy Trails*. Sidney, B.C.: Gray's Publishing Ltd.

Coleman, Loren. 2011. Wood knocking: More historical background. Online at http://www.cryptomundo.com/cryptozoo-news/woodknock-2; accessed December 10, 2013.

Gloia, Carol. 2011. Facts about the American Black Bear. Online at http://www.critters360.com/index.php./facts-about-the-american-black-bear-4012/; accessed December 18, 2013.

Gould, Rupert T. 1976. *The Loch Ness Monster and Others*. Secaucus, NJ: Citadel Press.

Green, John. 1973. *The Sasquatch File*. Agassiz, BC: Cheam Publishing Ltd.

———. 1978. *Sasquatch: The Apes Among Us*. Saanichton, B.C.: Hancock House.

Herrero, Stephen. 2002. *Bear Attacks*, rev. ed. Guilford, CT: The Lyons Press.

Kirk, John. 2006. Sasquatch in the Yukon. Online at http://www.cryptomundo.com/bigfoot/sasquatch-yukon; accessed December 9, 2013.

Kristian, Ken. 2013. Notes from my database on the Bushmen.
    Online at
    http://www.bigfootencounters.com/sbs/blackgiants.htm;
    accessed December 9, 2013.

Native American Bigfoot figures of myth and legend. 2013.
    Online at http://www.native-languages.org/legends-
    bigfoot.htm; accessed December 11, 2013.

Nickell, Joe. 1976. A "grin and bear it" story. *Whitehorse Star*
    (Yukon), June 21.

———. 1988. *Secrets of the Supernatural*. Amherst, NY:
    Prometheus Books.

———. 2008. Autobiographical Essay. *Contemporary Authors*
    269: 278–297.

———. 2012. *The Science of Ghosts*. Amherst, NY: Prometheus
    Books.

———. 2013. Bigfoot lookalikes, *Skeptical Inquirer* 37(4)
    (September/October): 12–15.

Patrick, Tom. 2009. Slow year for sasquatch sightings. Online
    at http://www.yukon-news.com/sports/slow-year-for-
    sasquatch-sightings; accessed December 10, 2013.

Sasquatch sightings reported in Yukon. 2005. CBC News (July
    13). Online at
    http://www.freedomcrowsnest.org/viewtopic.php?f=1&t
    =14641; accessed December 11, 2013.

Van Wormer, Joe. 1966. *The World of the Black Bear*. New
York: Lippincott.

Whitaker, John O., Jr. 1996. *National Audubon Society Field
Guide to North American Mammals*, rev. ed. New York:
Alfred A. Knopf

Searching for the Yowie: Australia's 'Great Hairy Man'

Like the fabled Yeti or Abominable Snowman of the Himalaya Mountains, and the Sasquatch/Bigfoot of North America, Australia's Yowie (or Yahoo, among many other names) is a supposed hairy man-beast that leaves strange tracks and wonderment wherever it ambles. Equated with an entity from Aboriginal mythology, also called *Dulagarl* (or *Doolagahl*, "great hairy man"), it was regarded as a magical being from the time of creation—what Aborigines call the Dreamtime. Interestingly, however, "[M]any early Europeans claimed to have seen the Yowie, many years before they came to learn about it from the aborigines" (Gilroy 1976, 9). It remains, according to cryptozoologist Loren Coleman (2006), "one of the world's greatest zoological or anthropological mysteries."

I first went in search of the creature in 2000 guided by skeptic Peter Rodgers. We ventured into the Blue Mountains west of Sydney, which—according to Yowie popularizer Rex Gilroy (1995, 212)—"continues to be a hotbed of Yowie man-beast activities—a vast region of hundreds of square miles containing inaccessible forest regions seldom if ever visited by Europeans." We drove into the Katoomba township bushland and took the world's steepest incline railway (originally built as a coal-mine transport in 1878) down into Jamison Valley rainforest where Gilroy himself once reported an encounter (1976, 10). We next drove to Jenolan Caves—which Gilroy (1995, 219) claims the Aborigines believed to be Yowie lairs—and bushwalked (hiked) through the surrounding mountainous terrain in a vain search for the elusive creature (Nickell 2001, 16–17).

In 2015, before and after the annual Australian Skeptics National Convention in Brisbane (of which I was honored to be a headliner) held October 16–18, I was able to resume my quest for the Yowie (and began several other investigations). I am indebted to Ross Balch for his tireless help in Brisbane and rural Queensland to the north, and to Kevin Davies and Nick

Ware for their dedicated assistance in Canberra and the New South Wales countryside. I photographed a Yowie with a rather wooden personality, did research at an Aboriginal institute, and kept an eye out for any exotic creature in the wild. Here is some of what I found.

Will the Real Yowie . . .

My study began when Ross Balch drove Myles Power and me through scattered Yowie territory to Yowie Park at Kilcoy. There, with parrots flitting about, we took photographs of the cracked wooden sculpture of the fabled man-beast (Figure 17.1). As I did so I quipped, "It doesn't get more real than this!" I meant that, of course, to apply to skeptics' sightings: Yowies seem not to appear to skeptics—even those looking for them, although it is obligatory for those who report encounters to insist they were previously skeptical.

But what about the Aboriginal elder who insisted, regarding the *Dulagarl*, "He only appears to Aboriginal people" (Mumbulla 1997)? Do the numerous non-Aboriginal sightings contradict him? Or is it possible he is talking about a quite different being—not the Yowie/Yahoo who today apes (so to speak) Bigfoot, but rather his people's supernatural entity who could induce sleep and fly through the air to kidnap lone women from the bush, yet who—according to some tribal/regional traditions—contrastingly carried clubs, used fire, and ate men. Other creatures of Aboriginal lore included the Quinkins who, variously shaped, were generally quite small; however, the giant Quinkin, Turramulli, towered over tall trees, had three taloned fingers on each hand, and as many clawed toes on each foot (Healy and Cropper 1994, 116, 118; 2006, 6–12). None of these entities sounds like a Bigfoot type.

Figure 17.1. Yowie statue in Yowie Park at Kilcoy, Queensland, Australia (Author's photo).

Indeed, just as Bigfoot was originally a "wild man of the woods" adapted from European tales and retrofitted onto Native American supernatural beings synthesized for the purpose (Nickell 2011; Loxton and Prothero 2013, 30–36), the Yowie/Yahoo is similarly derivative. Australian examples (from "A Catalogue of Cases" 2006) show that the earliest reports— the first in 1789 being acknowledged as "obviously a hoax," and continuing well after the beginning of the twentieth century—were sightings of a "WILD MAN or monstrous GIANT," a "Hairy Man," "in appearance half man, half baboon," "wild man of the bush," "like a blackfellow [Aborigine] only considerably larger," "hairy men," "an old man . . . covered with a thick coat of hair," "the hairy man of the wood," and so on.

As an example, in 1871 a "little girl" reported an encounter that was part of a "wild man" tradition in the area. She described an old man having a bent back, a covering of hair, tremendously long fingernails, and being about the height of her grandfather. He seemed to wish to avoid the girl ("A Catalogue" 2006, 207).

The term *Yowie* appears to have been used little if at all during this period, but the appellation "yourie" or "yowrie" appears by perhaps the 1920s, maybe after the Yowrie River or the nearby crossroads community of Yowrie, named by 1885. There was a Yowie Bay, but it was originally named Ewey Bay after the offspring of ewes called "eweys" (Healy and Cropper 2006, 13–143, 25), so the term *Yowie* may not be aboriginal at all ("A Catalogue" 2006, 217; Healy and Cropper 2006, 14, 25).

Yahoos

Moreover, if we consider the earlier, parallel term "Yahoo," we must at once recall that it was used—indeed invented[1]—by English satirist Jonathan Swift in his *Gulliver's Travels* (1726, 343–351). It describes his race of hairy, goat-bearded, manlike animals. Swift's Yahoos are brutes but, satirically, have human depravities.

By 1856 comes a report of a man-beast described alternately as a "wild man of the woods" or a "yahoo." A case of uncertain date in the 1860s involved a twelve-foot-tall Yahoo that had webbed feet and belched fire. In the same decade a Miss Derrincourt encountered "something in the shape of a very tall man, seemingly covered with a coat of hair . . . what the people here call a Yahoo or some such name." Still another case of that period involved an encounter with a "hideous yahoo" near an abandoned village ("A Catalogue" 2006, 204–206).

Graham Joyner (1994) conducted an in-depth study of the issue, which he reported in *Canberra Historical Journal*. He found that Aborigines probably adopted the term *Yahoo* from settlers, rather than the other way around.

'Littlefoot'

In addition to the "hairy giant" tradition, another type of Yowie is represented by the Aborigines' *Junjudees* (among other terms). These are small, hairy, magical creatures comparable to European fairies, elves, and leprechauns. Still, they seem as real as any if we believe the stories of teenagers who encountered them in 1978–1979 on Towers Hill, near Charters Towers, Queensland. One teen was attacked but claimed to have fought off the three-foot-tall creature with a rock (Healy and Cropper 2006, 120–121). Among many other reports were multiple sightings of similar creatures in 1994 in the vicinity of Carnarvon Gorge in Queensland (Pinkney 2003, 31–32).

Some Aborigines emphasize the Junjudees' supernatural powers, telling colorful tales about their exploits. For instance, they are guardian beings of certain places, are mischievous, and are attracted to honey. They are also a sort of bogeyman, used to keep children from wandering off, according to *Australian Folklore* (Ryan 2002, 137–138).

Yowie hunters, somewhat embarrassed by the little hairy folk, rationalize that they may be very young Yowies—no matter what Aborigines say about their "indigenous fairies" (as one researcher calls them [Povah 2006]).

What Manner of Beast?

Yowies are described in a widely diverse manner—beginning with height, which, based on 263 cases ("A Catalogue" 2006), ranges from about two to thirteen feet. The earliest-known record of an Aboriginal sighting came in July 1871 when a "gorilla"-like creature was encountered, but we must keep in mind that due to their long isolation on the Australian island–continent, the Aborigines had no knowledge of primates other than man. It was the early settlers and journalists who began to describe man-beasts with terms such as "huge monkey or baboon," "upright gorilla," and so on—from 1849 to the present.

In the early 1870s, in New South Wales, prospectors saw what they thought were "hairy men creeping around their tents," but a *Sydney Mail* correspondent concluded, "They were probably the large badgers or wombats which abound there" ("A Catalogue" 2006, 207). Wombats, marsupials that somewhat resemble (according to *American Heritage Dictionary* 1970) "small bears," may well be responsible for a number of other Yowie reports.

The kangaroo and its cousins the wallaby and the wallaroo are also good suspects. When, in 1954, three Queensland youths reported an encounter with a six-foot-tall creature covered with hair, possessing a long tail, and having an "apron" draped from its waist, the latter detail was an obvious clue pointing to a marsupial pouch. Someone suggested that the boys had been scared by "a cranky old wallaroo" ("A Catalogue" 2006, 224). Again, other cases may be explained by such related marsupials.

In addition to animals, there are numerous other possibilities: hoaxes, including those of a diminutive man who wore a hairy suit with bicycle-reflector eyes (Healy and Cropper 2006, 168–169); claims made by persons with fantasy-prone personalities and by ubiquitous attention seekers; real wild men, like a bearded, naked, mentally disturbed man mistaken for a Yowie ("A Catalogue" 2006, 262); and many other possibilities, including simple hallucinations and apparitional experiences. For example, "waking dreams" that occur between wakefulness and sleep (Nickell 2012, 353–354) may explain some cases of persons waking to see a Yowie looking at them (see Healy and Cropper 2006, 105, 170–171, 223). Again, like sightings of ghosts—typically seen when the percipient is tired, performing routine work, daydreaming, or the like—a Yowie's image may well up from the subconscious and be superimposed on the visual scene (Nickell 2012, 345).

'Bigfoot'

The American Sasquatch—after 1958 usually called Bigfoot (Nickell

2011, 68)—no doubt had an influence on Yowie sightings. That is especially so after 1967 when Roger Patterson's famous hoax film greatly publicized that elusive manimal (Nickell 2011, 68–72).

Patterson's "Bigsuit" (a modified gorilla costume) had pendulous breasts, one of several details found in Australian cases occurring (or being reported) only post-Patterson (at least as found in "A Catalogue" 2006). In addition to breasts, these motifs include Bigfoot's legendary foul odor; large, clawless, human-like footprints; and possibly other features (*cf.* Bord and Bord 2006, 215–310).

The Yowie is becoming increasingly standardized in its appearance. It is sometimes said that it resembles "depictions of the American Bigfoot" or that "America's Bigfoot would be an identical type" ("A Catalogue" 2006, 239, 271; *cf.* Nickell 2011, 225–229; 2013).

Even so, people I spoke with generally dismissed the Yowie. In 2000, for example, staffers at the information center at Echo Point in the Blue Mountains (Nickell 2001, 17) insisted the Yowie was only a mythical creature pursued by a few fringe enthusiasts, and this seems to remain the majority view. Several people laughed at my query, and a young bookstore employee in Canberra told me that, although having been born and raised in the Blue Mountains, she had never seen a Yowie or had reason to take the possibility seriously. Still, the wooden statue at Kilcoy stares vacantly on, a little monument to belief.

Note

1. The invention of the word is credited to Swift in *Gulliver's Travels*, 1726, by the authoritative *Oxford English Dictionary* (1971).

References

American Heritage Dictionary of the English Language. 1970. New York: Houghton-Mifflin Co.

Bord, Janet, and Colin Bord. 2006. *Bigfoot Casebook Updated*. N.p.: Pine Winds Press.

"A Catalogue of Cases 1789 to 2006." 2006. Appendix A of Healy and Cropper 2006, 203–295.

Coleman, Loren. 2006. Introduction to Healy and Cropper 2006, vii–viii.

Gilroy, Rex. 1976. *Psychic Australian*, August, 8–25.

———. 1995. *Mysterious Australia*. Mapleton, QL, Australia: Nexus Publishing.

Healy, Tony, and Paul Cropper. 1994. *Out of the Shadows: Mystery Animals of Australia*. Sydney: Ironbark.

———. 2006. *The Yowie: In Search of Australia's Bigfoot*. San Antonio, TX: Anomalist Books.

Joyner, Graham. 1994. Cited in Healy and Cropper 2006, 12–13.

Loxton, Daniel, and Donald R. Prothero. 2013. *Abominable Science!* New York: Columbia University Press.

Mumbulla, Percy. 1997. Quoted in Healy and Cropper 2006, 12.

Nickell, Joe. 2001. Mysterious Australia. *Skeptical Inquirer* 25:2 (March/April), 15–18.

———. 2011. *Tracking the Man-Beasts: Sasquatch, Vampires, Zombies, and More*. Amherst, NY: Prometheus Books.

———. 2012. *The Science of Ghosts*. Amherst, NY: Prometheus Books.

———. 2013. Bigfoot lookalikes. *Skeptical Inquirer* 37(5) (September/October): 12–15.

Oxford English Dictionary. 1971. Compact Edition, New York: Oxford University Press.

Pinkney, John. 2003. *Great Australian Mysteries*. Rowville, Victoria, Australia: The Five Mile Press.

Povah, Frank. 2006. Quoted in Healy and Cropper 2006, 123.

Ryan, J.S. 2002. The necessary other, or "when one needs a monster": The return of the Australian Yowie, *Australian Folklore* 17: 130–142.

Swift, Jonathan. 1726. Reprinted as *Gulliver's Travels into Several Remote Nations of the World*. Boston: Lee & Shepard, 1876.

Part 3:
A Menagerie of Cryptids, Strange Creatures, and Supernaturals

The New Zealand Moa: From Extinct Bird to Cryptid

On a visit to beautiful New Zealand in October 2015, I encountered this question: Did the moa, the large flightless, even wingless, bird of New Zealand—a cousin to the ostrich and the emu—really become extinct over 500 years ago? According to the historical and scientific evidence, the Maoris, who came in epic canoe voyages from Polynesia to settle the land in the thirteenth century ("Maori" 2016), drove them to extinction. Where once there were perhaps 58,000, by ca. 1440 there were none—due mostly to hunting (Figure 18.1) but also to forest clearing ("Moa" 2016). Some Maoris have even pointed to the legendary site—a Totara tree (an evergreen species) on the shore of Lake Rotorua—where their ancestors speared the last of what were once nine species of moa (Heuvelmans 1972, 134).

Figure 18.1. Artist's conception of a moa hunt, ca 1400 CE. (Drawing by Joe Nickell.)

'Obstinate Rumors'

Still, although Captain Cook made no mention of the colossal birds following his visit in 1769 (when he circumnavigated and charted New Zealand's two main islands), in time moa sightings would be reported, and they continue to the present. As fossilized bones and tracks piqued the interest of scientists beginning in 1839, there soon became "obstinate rumors that the moa survived." Indeed, when a German naturalist planned to climb Mount Taranaki (or Mount Egmont), a Maori chief dissuaded him, saying the mountain was guarded by a moa (Heuvelmans 1972, 137–138).

The reputed survival of the moa puts it in the category of "cryptid"— that is, a supposed animal that is "of interest to cryptozoology," the study of "hidden" animals. Cryptids are of two types: either (1) unknown species, such as Bigfoot or at one time the Mountain Gorilla; or (2) known species that supposedly become extinct but may have survived and could be rediscovered—the moa for instance (Coleman and Clark 1999, 75–77).

There were no fewer than 150 moa sightings beginning in 1500 (!) and increasing throughout much of the nineteenth century—in 1810, 1845, 1863, several circa 1825–1875, and one in 1878 by Sir George Grey. These are presented in the three-volume work *Moa Sightings* compiled by Bruce Spittle (2010), who is himself of part Maori descent. My old friend, the late skeptic Denis Dutton, professor of philosophy at the University of Canterbury in New Zealand, commented in the foreword: "This invaluable book ... provides a broad and systematic historical backdrop for individual claims to have sighted a living moa. However misguided or illusory moa-sighting claims may be, Bruce Spittle has done an authentic service to scholarship."

A Case in Point

The modern sightings begin (and Spittle's collection of reports ends) with an encounter by three hikers about fifty miles west of Christchurch on Wednesday, January 29, 1993. One witness was a local hotelier named Paddy Freaney; he saw and photographed the alleged bird, which he said stood about six feet tall (about half the height of the largest species of moa ["Moa" 2016]). Two others with him—Sam Waby, a school art department head, and Rochelle Rafferty, a gardener at Freaney's hotel—also insisted they saw the bird ("Long lost" 1993).

The possibility of hoaxing was soon raised. Dennis Dunbar, owner of the Moana Railway Station, told a reporter that he and Freaney had become friendly rivals in seeing who could attract the most publicity, and that months earlier Freaney had said his next exploit would be the best ever. However, Dunbar later retracted his statement (Coleman and Clark 1999, 164–167; Shuker 2003, 147–148).

Denis Dutton and the Canterbury skeptics invited the trio of witnesses to a meeting, held at the University Staff Club, to take their measure. They were treated in a friendly way—Denis would have made sure of that—and he felt that given "their coherence, consistency and sense of sincerity, the three were remarkable." He concluded that "few skeptics left the meeting thinking the moa sighting was an intentional hoax" as had been suggested (Dutton 1993).

An earlier plan by the Department of Conservation to search the immediate area for tracks, feathers, droppings, or any other sign of the moa ("Long lost" 1993) was nevertheless canceled when a then-postgraduate zoology student (later paleontologist), Richard Holdaway, identified the creature in the blurry photo as a red deer, observing that the neck was too thick for a bird ("Paddy" 2012).

Moa Tracks

Regarding tracks, Australian monster hunter Rex Gilroy—whose exploits I first encountered in 2000 (Nickell 2011, 91–94)—has repeatedly "discovered" moa tracks and other alleged traces in New Zealand.

Unfortunately, Gilroy is what skeptics disparagingly call a "repeater"—one who repeatedly claims to have had a most unusual sighting or to have made a remarkable discovery—thus straining credibility (Nickell 2011, 68, 215–216). For example, Gilroy alleges that in Australia's Blue Mountains (where I have also ventured), he encountered a Yowie, a Down Under version of Bigfoot (Nickell 2016). He also claims to have found Yowie tracks; sighted another presumed-extinct creature, the thylacine (or "Tasmanian Tiger"); and discovered evidence of such other cryptids as the Blue Mountain lion, the

Australian panther, and the megalania (an extinct giant monitor lizard) ("Rex and Moa" 2008).

In 1980, Gilroy claimed to have discovered a moa's lower leg bone in northern New Zealand. In 2001, he claimed he discovered no fewer than thirty-five separate moa ground prints from which he infers the existence of a colony of as many as fifteen birds. In November 2007, Gilroy reputedly found evidence of a moa nest (in a rotten, hollow tree trunk) and a trail of tracks through a forest. He stated that a cast of those prints could be matched with a species known as the little scrub moa that he had viewed in the Auckland Museum ("Rex and Moa" 2008). Gilroy has also exhibited what he maintained was a fossilized footprint of the extinct giant ape *Gigantopithecus*, which would make it the only such print known (Healy and Cropper 2006, 50).

Gilroy withholds information regarding some of his evidence, and that raises further questions. For instance, he refused to reveal the site of his alleged 2007 discoveries or to allow anyone, even a television cameraman, to visit the secret location. He says that his intent is to protect the birds, but secrecy can invite suspicion. New Zealand paleontologist Joan Wiffern—the first to discover dinosaur bones in the country—suggested Gilroy was perhaps only enjoying "a bit of a dream" ("Rex and Moa" 2008).

Figure 18.2. Author poses with moa, reconstructed from skeleton, in Auckland museum. (Author's photo by Craig Shearer.)

Explanations

Given the lack of any real evidence for the supposed moa survival, a moa expert at Canterbury Museum, Beverly McCulloch, stated: "The weight of scientific evidence is against it. The history of moa studies is littered with possible sightings, none of which has ever been proven, some of which were hoaxes and most of which were wishful thinking" ("Long lost" 1993). Wishful thinking could cause people to misperceive (as with reports of Bigfoot that may often be misidentifications of bears walking upright [Nickell 2013]). In the

176

1993 case, for example, some thought the "moa" might have been an emu, since the birds are bred on nearby farms, but local farmers insisted theirs were accounted for ("Long lost" 1993; Shuker 2003, 143–150). Nevertheless, emus do escape, and they can be mistaken for moas when they do ("Baby moa" 2013).

In addition to hoaxes and misidentifications, belief in the continued existence of the moa may also have been assisted by the dynamics of folklore. Oral tradition tends to "telescope" time so that important events are brought into more direct association (Layton 1999, 25). Thus, it appears that the last moa hunt (previously mentioned) "was constantly updated to 'grandfather's time' to keep it in touch with the present" ("Review" 2010). In this way stories of ancient moa encounters might seem historically much more recent and could help inspire belief that sparks more claims.

References

Baby moa. 2013. Available online at
http://www.stuff.co.nz/oddstuff/8983326/Baby-moa-
spotted-in-woman's-garden; accessed March 2, 2016.

Coleman, Loren, and Jerome Clark. 1999. *Cryptozoology A to Z: The Encyclopedia of Loch Monsters, Sasquatch, Chupacabras, and Other Authentic Mysteries of Nature*. New York: Fireside/Simon & Schuster.

Dutton, Denis. 1993. Skeptics meet moa spotters. *New Zealand Skeptic*. 30 (December): 1.

Gazin-Schwartz, Amy, and Cornelius J. Holtorf. 1999. *Archaeology and Folklore*. New York: Routledge.

Healy, Tony, and Paul Cropper. 2006. *The Yowie: In Search of Australia's Bigfoot*. New York: Anomalist Books.

Heuvelmans, Bernard. 1972. *On the Track of Unknown Animals*. Cambridge, MA: The MIT Press, 131–146.

Layton, Robert. 1999. In Gazin-Schwartz and Holtorf, 1999.

Long lost bird 'seen by hikers.' 1993. *Christchurch Press*. Undated clipping (January 21–26). Available online at http://forteanworld.tumblr.com/post/74367840405/artic les-moa; accessed February 23, 2016.

Maori. 2016. Available online at http://www.maiori.com/; accessed February 23, 2016.

Moa. 2016. Available online at
https://en.wikipedia.org/wiki/Moa; accessed February
22, 2016.

Nickell, Joe. 2011. *Tracking the Man-Beasts*. Amherst, NY:
Prometheus Books.

———. 2013. Bigfoot lookalikes: Tracking hairy man-beasts.
Skeptical Inquirer 37(5) (September/October): 12–15.

———. 2016. Searching for the Yowie, the Down Under Bigfoot.
Skeptical Inquirer 40(2) (March/April): 16–18.

Paddy Freaney, Moa man, is dead. 2012. Available online at
http://kiwiscots.blogspot.com/2012/03/paddy-freaney-
moa-man-is-dead.html; accessed February 24, 2016.

Review: Moa sightings. 2010. Available online at
http://www.strangehistory.net/2010/12/22/review-moa-
sightings; accessed February 23, 2016.

Rex and moa articles. 2008. Available online at
http://www.network54.com/Forum/93258thread/
1201218041/last-1201218183/Rex+and+Moa+articles;
accessed March 1, 2016.

Shuker, Karl P.N. 2003. *The Beasts That Hide from Man*. New
York: Paraview Press.

Spittle, Bruce. 2010. *Moa Sightings*. Dunnedin, NZ: Paua Press
Ltd. Cited in Review 2010.

The Giant Panda Discovered in the Land of Myth

Its immense popularity today belies the fact that the panda was once among the world's most obscure creatures, "as mythical and elusive as Bigfoot" (Edwards 2009). Bigfooters are prone to emphasizing such creatures that were only discovered comparatively recently—for example a giraffe relative, the okapi (1901), and a "living fossil" fish, the coelacanth (1938)—because they "symbolize the search for Bigfoot is not over" (Edwards 2009). Inspired by my encounter with pandas during a trip to China in 2010 as a visiting scholar (see Figure 19.1), I have since looked into their fascinating history.

Legendary Creature

In ancient China, the panda was an exotic creature—rare, even mythic (like the dragon). Texts from very ancient times describe a lumbering, black-and-white animal believed to have been a panda.

Figure 19.1. "Self-portrait with panda" at Giant Panda House, Beijing Zoo, 2010. (Author's sketch)

The Dowager Empress Bo was reportedly interred in her tomb (ca. 170 BCE) with a panda skull—whether as treasure or talisman, or both, is unclear (Schaller 1994, 61–62). Also, ancient poetry tells of the gift of a pelt that may well have been from a panda ("Pandas" 2017). Such pelts' distinctive appearance and rarity gave them great value—not to mention alleged magical properties. According to the earliest Chinese "encyclopedia" (or reference book), *Erya*, dating from the Qin dynasty (221–207 BCE), sleeping on panda fur supposedly regulated a woman's menstrual cycle. The later poet Bai Juyi (772–846 CE) attributed to the pelts both curative properties and the power to exorcize evil spirits.

A record from 210 CE reports an emperor in the ancient capital of Xian keeping several of the remarkable creatures as pets. Later, in the Tang dynasty (618–907 CE), China sent a goodwill gift to the Japanese emperor, consisting of a pair of pandas in addition to panda pelts.

Again, in ancient times, lack of familiarity with the rather placid creatures caused many Chinese people to fear them, thinking they were monsters. They were described as black-and-white "tapirs" (herbivorous mammals with short legs). The *Erya* and other ancient books said the animals had a propensity for eating metal—copper and iron. (However, this may have been due to pandas sometimes entering villages where they would lick and chew tasty cooking pots. Their powerful jaws enable them to crunch metal pots, and one panda that munched on its water pan later defecated aluminum bits) (Olesen 2014; Schaller 1994, 61).

Despite all such mentions, however, the absence of depictions of pandas in Chinese art before the twentieth century seems to confirm their relative obscurity throughout the country's long history (Schaller 1994, 10–11, 61–62; "Giant panda" 2017; Olesen 2014; "Pandas" 2017).

Quest for a Live Panda

The obscurity was even greater in the West, and scientists had no sure knowledge of pandas. Then, between 1865 and 1869, the French missionary and naturalist Armand David (1826–1900) crossed eastern Asia. On March 11 of the latter year—having already discovered two new mammals (a species of deer and a "snow monkey")—he made his third major discovery. While taking tea with a wealthy Sichuan landowner, he noticed a fur pelt he recognized as that of a legendary animal he had previously heard of, "the famous black-and-white bear." (His story is told by Bernard Heuvelmans [1972, 14–16], who is known as "the father of cryptozoology" [Nickell 2016, 35–36].)

David was now persuaded that it was a real creature, and he commissioned hunters to bring him one. In just twelve days they did, actually capturing a young one, but having to kill it in order to transport it. Thus, the naturalist became the first westerner to secure a specimen fur of what is now known as the giant panda. Subsequently the hunters brought him an adult's skin and skeleton and a few days later a specimen of an already known creature: the kitten-sized red panda (Heuvelmans 1972, 14–17).

It was not until 1914—almost half a century after David's discovery of the giant panda for science—that a westerner actually saw a live panda in the wild. That honor fell to German zoologist Hugo Weingold, and soon museums launched expeditions seeking specimens for their collections (Coleman and Clark 1999, 92).

President Theodore Roosevelt's sons, Theodore Jr. and Kermit, became the first to collect a panda specimen for an American museum, Chicago's Field Museum of Natural History. On two expeditions, in 1925 and 1928, the Roosevelts, accompanied by trained naturalists, obtained thousands of specimens of birds, reptiles, and mammals. Reportedly the brothers simultaneously shot their panda out of a tree. Its skin and a second one obtained by a local hunter were expertly stuffed and displayed in the museum's new Asian Hall in 1929. I saw them in a beautiful diorama there in 1982 ("Field Museum" N.d.).

It remained for the first live giant panda to be captured and brought out of its mountain lair. An expedition to accomplish that was led by adventurer William H. Harkness in 1933–1934, but it failed due to the politics of obtaining a permit as well as instability in the region. After Harkness died in Shanghai in February 1936, his widow, Ruth Harkness (1900–1947), traveled to China by boat and in July took over the expedition. Luckily, she discovered a cub in the Wassu region. Assisted by the expedition's Yang Di Lin and the later Yeti hunter W.M. "Gerald" Russell, she transported "Su-Lin" to the Brookfield Zoo in Chicago in 1937. (On a subsequent expedition, she brought back two additional giant pandas [Coleman and Clark 1999, 104–105, 209].)

There is more to the story of Su-Lin. According to Heuvelmans (1972, 29), Mrs. Harkness had found the cub in a hollow tree. "It was crying as if its heart was fit to break. She picked up this child of her husband's dreams and nursed it in her arms." In December 1936, having "fallen in love as she bottle-fed Su-Lin," reports *National Geographic*, she boarded a ship at Shanghai for her return voyage, carrying a wicker basket. No doubt remembering her husband's trouble getting an export permit, she proffered one that read, "One dog, $20.00." Thus, Ruth Harkness—erstwhile socialite and clothing designer, turned adventurer with a can-do attitude—brought to the modern world a living panda (Holland 2017).[1]

Pandamania

It is not true, as one source reports ("Pandas" 2017), that "The word *pandemonium* was coined in 1936 to describe the reception a panda [Su-Lin] received when it was first shown in the West." *Pandemonium* was in fact first used by the poet John Milton in his 1667 *Paradise Lost:*

A solemn Council forthwith

To be held at Pandaemonium,

The high capital of Satan and his Peers.

Milton coupled the Greek *pan* ("all") with the existing Latin word *demonium* ("abode of demons," i.e., hell); thus *pandemonium* is "the place of demons."

Nonetheless, on the arrival of Su-Lin at Brookfield in 1936, what is now well described as *pandamania* first occurred. On opening day of the exhibit, over 53,000 visitors appeared. The mania has continued. In what has become known as "panda diplomacy," the Chinese government revived a policy that dates back to the previously mentioned gift to the Japanese emperor.

During the years 1958 to 1982, the People's Republic of China gave a total of twenty-three of the beloved creatures to nine countries (but by 1984 had amended offerings to ten-year loans). After President Nixon's historic visit to the country in 1972, China gave a pair of pandas, Ling-Ling and Hsing-Hsing, to the United States—again with a tremendous reaction. First Lady Pat Nixon held a welcoming ceremony at the National Zoo, drawing more than 20,000 visitors. During the first year, the pandas drew an estimated 1.1 million viewers (Holland 2017; "Panda Diplomacy" 2017).

But What Is a Panda?

Having gotten a panda, science had considerable trouble knowing what to do with it—in terms of classification, that is.

Early references to the creature were scant. Armand David had called it, obviously after others, "the famous black-and-white bear." It had also been termed the "bamboo-bear," while some, with reference to the missionary himself, now called it "Pére [Father] David's bear."

However, when mammalogist Alphonse Milne-Edwards examined a skeleton and studied the animal's dentition and bone structure, he realized, says Heuvelmans (1972, 16), "with a touch of genius that it was related to the Procyonidae, the raccoons," and gave it the name of *Ailuropoda melanoleucus*, "the black-and-white cat-foot."[2] (Its traditional Chinese name is "big bear cat.")

In fact, the giant panda (so-called to distinguish it from the previously known "panda," the red panda, only a very distant relative), shares features with both raccoons and bears. But molecular genetic studies of bears (the family Ursidae) now show that the giant panda is, after all, a true bear, although early in history it branched from the bear family tree. So those of us who were once corrected for using the name "panda bear" may now do so freely (Fergus 2005).

As a Cryptid

But was the panda bear ever "as mythical and elusive as Bigfoot," as quoted at the beginning of this report? Not quite. As a Listverse site acknowledges, "The existence of the giant panda has never been disputed by the scientific community"—not as a whole, that is.[3] Thus, the comparison of the panda with Bigfoot is erroneous in that respect, since Bigfoot's existence is nearly universally doubted by scientists and science-based investigators.

However, the Listverse site goes on to add that, "therefore, it [the giant panda] has never been a true cryptid" ("Top 10 Cryptids" 2010). But that is a mistaken interpretation of what it means to be a cryptid. As Coleman and Clark (1999, 15) observe:

> Heuvelmans prefers "hidden" to the "unknown" because to those people who live near them, the animals are not unfamiliar; if they were, there would be no native accounts, and we would never have heard of them. They are, however, undetected by those who would formally recognize and catalogue them.

So like Bigfoot, "the famous black-and-white bear" was indeed a cryptid, according to cryptozoologists' use of the term.

On the other hand, there are important differences between the panda and Bigfoot as to cryptid status. So far as we know, no one ever gave a pair of Bigfoot or Yeti creatures as a gift. Their pelts were not obtained and used as magical objects. (Some alleged "Yeti fur" turned out to have belonged to the rare Tibetan blue bear, and a "Yeti scalp" came from the serow, a goatlike animal [Nickell 2011, 61].) Naturalist missionary David obtained a panda specimen just days after learning of it, and other specimens—complete skins, skeletons, and, in time, taxidermied then living animals—were eventually

185

displayed. Yet Bigfoot-type creatures still lack any credible evidence since 1811 when a track (almost certainly a grizzly's) was reportedly seen in the Alberta Rockies. There were few if any panda hoaxes, but Bigfoot ones are common, including Roger Patterson's 1967 film of "Bigsuit" (Nickell 2011, 66–73).

Most importantly there is the matter of fossil evidence. The skull of a "pigmy-sized" giant panda—the latter's earliest-known ancestor, some two million years old—was discovered in a south China cave ("Remains" 2007). Fossil evidence shows that—while the panda bear is now found only in a limited region—it once was widespread in China (Fergus 2005). In contrast, no fossil evidence in North America or elsewhere has been found for the legendary man-beast. Yet, often imitated by an upright-standing bear, it has taken on a life, so to speak, of its own—though never really *real*, like the "black-and-white bear" the world has come to love.

Notes

1. I have been unable to confirm a rumor that Mrs. Harkness actually purchased the cub from a hunter, who acted for a rival would-be museum supplier ("Pandas" 2017). Indeed, Harkness seems supported by the evidence (Schaller 1994, 49). See also Croke, 2006.

2. The name is from Greek *ailouros* ("domestic cat") and *poda* ("foot") together with ancient Greek *melano* ("black") and *leukos* ("white").

3. In his book, *All the Presidents' Children*, Doug Wead (2003, 199) briefly mentions the "many naturalists who had doubted [the giant panda's] existence."

References

Coleman, Loren, and Jerome Clark. 1999. *Cryptozoology A to Z*. New York: Simon & Schuster.

Croke, Vicki Constantine. 2006. *The Lady and the Panda*. New York: Random House.

Edwards, Guy. 2009. Panda discovered in 1927 was once as elusive as Bigfoot. Available online at http://www.bigfootlunchclub.com/2009/11/ panda-discovered-in-1927-was-once-as.html; accessed December 30, 2016.

Fergus, Charles. 2005. *Bears: Wild Guide*. Mechanicsburg, PA: Stackpole Books.

The Field Museum: Theodore and Kermit Roosevelt. N.d. Available online at http://www.fieldmuseum.org/theodore-and-kermit-roosevelt; accessed January 6, 2017.

Giant panda. 2017. Available online at https:/en.wikipedia.org/wiki/Giant-panda; accessed January 9, 2017.

Heuvelmans, Bernard. 1972. *On the Track of Unknown Animal*. Cambridge, MA: The MIT Press.

Holland, Jennifer S. 2017. Who Discovered the Panda? Available
online at
http://www.nationalgeographic.com/magazine/2016/08/
explore-panda-mania-history; accessed January 6,
2017.

Nickell, Joe. 2011. *Tracking the Man-Beasts*. Amherst, NY:
Prometheus Books.

———. 2016. Creators of the paranormal. *Skeptical Inquirer*
40(3) (May/June): 32–39.

Olesen, Alexa. 2014. Chinese People Used to Think Pandas
Were Monsters. Available online at
http://foreignpolicy.com/2014/10/23/chinese-people-
used-to-think-pandas-were-monsters; accessed January
4, 2017.

Panda Diplomacy. 2017. Available online at
https://en.wikipedia.org/wiki/Panda_diplomacy;

accessed January 4, 2017.Pandas. 2017. Available online at
http://factsand-
details.com/china/cat10/sub68/item379.html; accessed
January 5, 2017.

Remains of earliest giant panda discovered. 2007. *Science Daily*
(June 19). Available online at www.sciencedaily.com/
releases/2007/06/070618174710.htm; accessed
January 26, 2017.

Schaller, George B. 1994. *The Last Panda*. Chicago: U of
Chicago Press.

Top 10 Cryptids That Turned Out to be Real. 2010. Available online at http://listverse.com/2010/08/13/top-10-cryptids-that-turned-out-to-be-real; accessed December 30, 2016.

Wead, Doug. 2003. *All the Presidents' Children*. New York: Atria Books.

Solving 'Mothman' (and Others)

The scary movie, *The Mothman Prophecies* (2002) tells the story of a reporter (played by Richard Gere) who is drawn to a West Virginia town by eyewitness accounts of a flying monster. From November 1966 to November 1967, residents in the vicinity of Point Pleasant (near the Ohio state line) were frightened by "Mothman" (whose appellation was a reporter's takeoff on the then-current *Batman* TV series). The movie is based on a book of the same title by arch paranormal mystery monger John A. Keel (1975). Keel rounded up giant bird reports, both local and worldwide, and combined them with UFO sightings, visits by Men in Black, telephone predictions from alleged extraterrestrials and their "contactees" (precursors of the "abductees"), plus a tragic bridge collapse and sundry other elements.

Figure 20.1. *The Mothman Prophecies* movie poster (Screen Gems / Sony 2002)

Background

"Mothman" was encountered one night about seven miles from town when two couples drove through an abandoned complex popularly called the TNT area (after its World War II use for making munitions). About 11:30 p.m. they saw the shining red eyes of a creature, "shaped like a man, but bigger," one witness would say. "And it had big wings folded against its back." It was further described as greyish and walking on sturdy legs with a shuffling gait (Scarberry 1966). As it took flight and seemed to follow them, it "wasn't even flapping its wings" but "squeaked like a big mouse" (quoted in Keel 1975, 52–53).

Soon others were seeing the winged enigma, including two Point Pleasant firemen who visited the TNT area just three nights after the couple's sighting. They too saw the red eyes and described the creature as "huge" but were emphatic: "It was definitely a bird" (Keel 1975, 56). Most reports described it as headless yet with large, shining red eyes set near the top of its body. Not all accounts agreed, however: One woman stated that what she saw "had a funny little face" although she "didn't see any beak," just those "big red poppy eyes." Keel also describes some "gigantic birds" about seventy miles to the north, in Ohio, that had a ten-foot wingspan and heads with "a reddish cast," yet lacking "the famous glowing red eyes" (Keel 1975, 60–61).

Except for an exaggeration of size—perhaps caused by an overestimate of the intervening distance—the Ohio birds seem to fit the appearance of the common turkey vulture which can have a six-foot wingspan and an unfeathered red head.

Eyeshine

The reflector-like nature of the creature's eyes is revealing. As ornithologists well know, some birds' eyes shine bright red at night when caught in a beam from auto headlights or a flashlight. "This 'eyeshine' is not the iris color," explains an authority, "but that of the vascular membrane—the tapetum—showing through the translucent pigment layer on the surface of the retina" (Gill 1994).

The TNT area, which I visited both days and nights, is surrounded by the McClintic Wildlife Management Area—then, as now, a bird sanctuary! Owls, which exhibit eyeshine, populate the area. Indeed, Steve Warner (2002), who works for West Virginia Munitions to produce .50-caliber ammunition in the TNT compound, told me there were "owls all over this place." Conversely, neither he nor a coworker, Duane Chatworthy (2002), had ever seen Mothman, although Warner pointed out he had lived in the region all of his life.

Because of Mothman's squeaky cry, "funny little face," and other features, including its presence near barns and abandoned buildings, I first identified it as the common barn owl (Nickell 2002). One *Skeptical Inquirer* reader (Long 2002) insisted it was instead a great horned owl which, although not matching certain features so well, does have the advantage of larger size. It seems likely that various owls and even other large birds played Mothman on occasion.

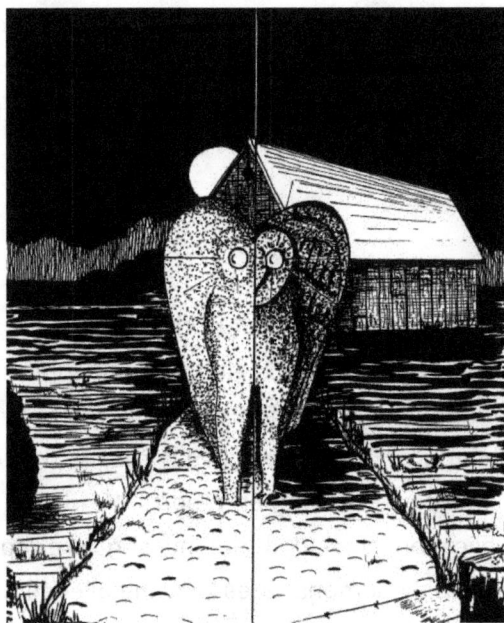

Figure 20.2. Mothman (left) compared to owl. (Drawing by Joe Nickell.)

I did some further research regarding eyeshine, learning that the barn owl's was "weak" and the great horned owl's only "medium." However the *barred* owl exhibits "strong" eyeshine (Walker 1974) and—according to David McClung (2002), wildlife manager at McClintic—is common to the area; indeed, it is even more prevalent there than the barn owl. It is also larger than the barn owl, which it somewhat resembles, and is "only a little smaller than the Great Horned Owl" (Kaufman 1996, 317). (Mounted specimens of these and other species of owls are profusely displayed in the West Virginia State Farm Museum near the McClintic preserve. Museum Director Lloyd Akers generously allowed me special access to examine and photograph them.)

In light of the evidence it seems very likely that the original Mothman was a barred owl and that other sightings were mostly caused by owls—probably more than one type. A man named Asa Henry shot and killed one, tentatively identified as a snowy owl, during the Mothman flap. Although only about two feet tall, a newspaper dubbed it a "giant owl" due to its wingspan of nearly five feet (Sergent and Wamsley 2002; 94, 99). In Point Pleasant I was able to view the mounted specimen and to speak with Mr. Henry's grandson, David Pyles. Himself a taxidermist, Pyles (2002), who is "very skeptical of Mothman," told me his grandfather always maintained that the Mothman flap ended after he had shot the bird.

"Bighoot"

Owls are very likely responsible for other birdman sightings. One of these is the 1952 case of the Flatwoods Monster that supposedly arrived in Flatwoods, West Virginia, aboard a flying saucer. Loren Coleman in his Mothman and Other Curious Encounters (2002) sees in that case "elements foreshadowing" the subsequent Mothman reports. However, as a Michigan Audubon Society publication concluded, my investigative report on the case (Nickell 2000) "convincingly demonstrates that the alleged flying saucer was really a meteor and the hissing creature from outer space was none other than a Barn Owl! Check it out, it's a real scream!" (Those Monster Owls 2001).

194

Again, in 1955, there was the "nighttime attack" of so-called "little green men" at a farmhouse near Kelly, Kentucky. Actually, the eyewitnesses described the creatures as "silver" and later downsized them to "two and a half feet tall." They had large pointy ears, shining yellow eyes, clawlike "hands," and "spindly" legs—features that helped identify them—no more than two ever being seen at one time. At the event's fiftieth anniversary festival in nearby Hopkinsville (where I received from the mayor a key to the city), I offered my identification of a pair of territorial Great Horned Owls (see Nickell 2011, 167–173).

Somewhat similarly, several sightings in 1976 in Cornwall, England, featured a "big feathered bird man" that was first seen "hovering over a church tower"—a common nesting place for barn owls (Kaufman 1996, 306). Appropriately, the entity became known as "Owlman" (Coleman 2002, 34–36).

As to Mothman, "cryptozoologist" Mark A. Hall (1998) has opined that it may be a hitherto undiscovered species of giant owl! He has dubbed it "Bighoot" and cites evidence that it has long existed in the Point Pleasant area. I take this as an implicit concession that Mothman—of all the creatures known to science—most resembles a barred owl, except for size.

Here then is the question separating the mystifiers from the skeptics: Is it more likely that there has long been a previously undiscovered giant species among the order strigiformes (owls), or that some people suddenly encountering a "monster" at night have misjudged its size? The latter possibility is supported by the principle of Occam's razor, that the simplest tenable explanation is to be preferred as most likely correct. The principle seems especially applicable to the case of Mothman.

References

Chatworthy, Duane. 2002. Interview by author, April 12.

Coleman, Loren. 2002. *Mothman and Other Curious Encounters*. New York: Paraview Press.

Gill, Frank B. 1994. *Ornithology*, 2nd ed. New York: W. H. Freeman and Co., 188.

Hall, Mark A. 1998. Bighoot—the giant owl. *Wonders* 5:3 (September), 67–79; cited in Coleman 2002.

Kaufman, Kenn. 1996. *Lives of North American Birds*. New York: Houghton Mifflin Company.

Keel, John A. 1975. *The Mothman Prophecies*. Reprinted New York: Tor, 1991.

Long, Chris. 2002. Letter to editor, *Skeptical Inquirer*, July/August, 66.

McClung, David. 2002. Interview by author, April 13.

Nickell, Joe. 2000. The Flatwoods UFO monster. *Skeptical Inquirer*, November/December, 15–19.

———. 2002. 'Mothman' Solved! *Skeptical Inquirer*, March/April, 20–21.

———. 2011. *Tracking the Man-Beasts*. Amherst, NY: Prometheus Books.

Pyles, David. 2002. Interview by author, April 12.

Scarberry, Linda. 1966. Handwritten account reproduced in
    Sergent and Wamsley 2002, 36–59.

Sergent, Dorrie, Jr., and Jeff Wamsley. 2002. *Mothman: The
    Facts Behind the Legend*. Point Pleasant, W. Va.:
    Mothman Lives Publishing.

Those monster owls. 2001. *The Jack-Pine Warbler*, March/April,
    6.

Walker, Lewis Wayne, 1974. *The Book of Owls*. New York:
    Alfred A. Knopf, 218–222.

Warner, Steve. 2002. Interview by author, April 12.

Montauk Monster and the
Raccoon Body Farm

In July 2008, the carcass of a creature soon dubbed the "Montauk Monster" allegedly washed ashore near Montauk, Long Island, New York (Figure 21.1). It sparked much speculation and controversy, with some suggesting it was a shell-less sea turtle, a dog or other canid, a sheep, or a rodent—or even a latex fake or possible mutation experiment from the nearby Plum Island Animal Disease Center. (In time, other "Montauk Monsters" turned up—one, for example, a decomposing cat [Naish 2008].)

Before long, the original creature was credibly identified as a raccoon by wildlife biologist Jeff Corwin (Boyd 2008). Although questions remained, I gave the matter little more attention—for a time.

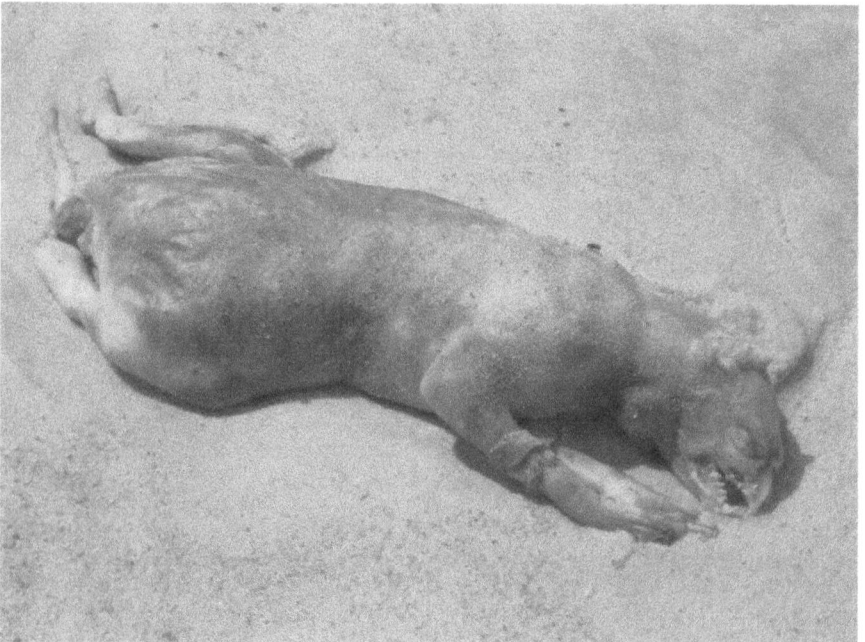

Figure 21.1. This photo of the Montauk Monster was widely circulated on the Internet, causing much speculation.

Case of the Missing Hair

However, when—on an investigative outing on September 19, 2009—I came across a dead raccoon by the roadside, I quickly decided it might be profitable to study the issue further. My wife, Diana, drove the getaway car while I retrieved the roadkill in busy traffic. I subsequently deposited it at a convenient wooded site she dubbed the Raccoon Body Farm (after the famous forensic site maintained by the University of Tennessee). (See Figure 21.2.) I monitored it to observe developments.

Figure 21.2. Raccoon roadkill is studied at the author's Raccoon Body Farm. (Photo by Joe Nickell)

The experiment raised questions. The already putrid carcass decomposed quickly, and in about three days it was largely gone, leaving behind a swarming mass of maggots plus *all of the raccoon's fur* (Figure 21.3). As I looked again at the Montauk Monster photo, I thought the creature's fur loss needed explaining. One suggestion was mange (Radford 2009), which can produce strange-looking creatures. (Indeed, Diana and I once went in search of a Bigfoot in Pennsylvania that turned out to be a mangy bear [Nickell 2008;

199

"Big Foot" 2008]. More recently I examined, up close, a mangy coyote mistaken for a "*chupacabra*" near Springfield, Missouri [Nickell 2011].) However, long familiar with mange from my boyhood days in eastern Kentucky, I did not think the Montauk Monster looked like a case of mange.

Figure 21.3. After three days, the decomposition is advanced, but the animal's fur remains. (Photo by Joe Nickell)

Paleontologist and science blogger Darren Naish (2008) observed that water-logged creatures often lose their fur. But what was a raccoon doing in the ocean in the first place—if it did not just die on the beach, and if it really was a raccoon?

A housefly on the creature's back allowed photo enlargement (by colleague Tom Flynn) to be made approximately life size (assuming an upper limit for the fly as 12mm) and the carcass to be measured as about 65cm (approximately 25.6 inches) long. This is well within the range of the adult common raccoon, *Procyon lotor* (according to the National Audubon Society's *Field Guide to North American Mammals* [Whitaker 1996, 748], which gives a length range of 24–37 inches).

Those who doubted the raccoon identification had their main

200

arguments refuted by Darren Naish (2008). First, whereas the creature was said to be too long-legged for a raccoon, Naish observed: "Raccoons are actually surprisingly leggy"; he asserted that "claims that the limb proportions of the Montauk carcass are unlike those of raccoons are not correct." Secondly, claims that the creature had a "beak" prompted Naish to say of raccoons: "The tendency for the soft tissues of the snout to be lost early on in decomposition immediately indicates that the 'beak' is just a defleshed snout region: we're actually seeing the naked premaxillary bones. . . . The Montauk animal has lost its upper canines and incisors (you can even see the empty sockets [in one photo]). . . ."

Viking Funeral?

I recalled an earlier claim that the presence of a hairless raccoon at Montauk had been explained—and then the explanation dismissed as not credible. Reportedly, three young men had found a dead raccoon on nearby Shelter Island two weeks earlier. As a lark, they gave it a "Viking funeral": sending it adrift on a makeshift raft (made of twigs and an inflatable toy)—containing a watermelon and cloth scraps—after setting the carcass afire. (Their prior revelry involved a "waterboarding endurance competition," and later hijinks included a "clothespins-on-your-genitals challenge." Many were skeptical of the trio's story, pointing out what a circuitous fifteen-mile route the carcass would have had to travel to get to Montauk ("The Latest" 2009).

However, an investigator is not a dismisser who ignores evidence because it is inconvenient or merely because someone's behavior does not comport with what he or she thinks someone would do in a situation. Neither is an investigator the equivalent of a newspaper's rewrite staffer. Mysteries are solved by the use of the best, corroborative evidence, together with the principle of Occam's razor (that the preferred hypothesis is the one that makes the fewest assumptions consistent with the evidence). It turns out there is considerable corroborative evidence for the "Viking funeral" claim.

First, data on the surface currents and winds in the area show that the "Viking funeral" critter would likely have been pushed in the proper direction ("The Latest" 2009). Significantly, the trio provided photographs documenting their launching. The snapshots (see Figure 21.4) clearly show a dead raccoon—first being launched on a raft of sticks as claimed, then blazing and adrift. Also the Montauk Monster has what appears to be a strip of cloth around its right foreleg, possibly linking it to scraps of cloth used with the "Viking funeral" raccoon. (Enlargement of one of the trio's photos shows what could be a band around the raccoon's right foreleg.) Moreover, the forelegs of the latter are in the same approximate position with respect to each other as those of the Montauk Monster ("Has the Montauk" 2009). Finally, the latter's flesh has a decidedly baked appearance, consistent with the reported burning.

Figure 21.4. A "Viking funeral" appears to account for the presence and condition of the Montauk Monster.

Therefore, the best evidence thus far indicates—until perhaps better evidence comes to light—that the Montauk Monster was neither a hoax (involving either a fake latex creature or a skinned animal) nor a mangy, gone-swimming-and-drowned critter; instead, it is an identifiable raccoon whose dead body was set ablaze and adrift on a makeshift raft as part of a

202

comically wry ritual dubbed a "Viking funeral." The dead raccoon does seem to be achieving a kind of immortality as a result.

References

Big Foot in the Pennsylvania wilds. 2008. Online at http://www.angelfire.com/pa2/stonemanguitars/bigfoot. html; accessed February 27, 2008.

Boyd, Aaron. 2008. Naturalists confirm Montauk Monster is relative of Rocky Raccoon. Online at http://www.hamptons.com/print.php?articleID=4474; accessed December 23, 2009.

Has the Montauk Monster mystery been solved? 2009. Online at http://gawker.com/5278112/has-the-montauk-monster-mystery-been-solved; publ. June 4, 2009; accessed December 22, 2009.

The latest Montauk Monster theory: A compleat accounting. 2009. Online at http://gawker.com/5280493/the-latest-montauk-monster-theory-a-com; accessed December 22, 2009.

Naish, Darren. 2008. What was the Montauk monster? Online at http://scienceblogs.com/tetrapodzoology/2008/08/the_montauk_monster.php; accessed October 27, 2009.

Nickell, Joe. 2008. Personal journal entry, February 24.

———. 2011. Chupacabra attack (blog post). Available online at http://www.centerforinquiry.net/blogs/entry/Chupacabr a_attack/; accessed February 17, 2012.

Radford, Benjamin. 2009. Hide the kids and wake the neighbors: The Montauk Monster returns! *Skeptical Briefs* 19(3) (September): 14.

Whitaker, John O. Jr. 1996. *Field Guide to North American Mammals*, revised ed. New York: Alfred A. Knopf.

Song of a Siren: A Study in Fakelore

During an investigative tour of Germany in 2002 (Nickell 2003), I explored along the beautiful Rhine Valley guided by my Center for Inquiry–Germany colleague Martin Mahner. There, we tracked a headless ghost (Nickell 2012, 33–34) and viewed the lair of the beautiful, enchanting Lorelei (associated with a massive rock 430 feet high, near St. Goar [Zieman 2000]). And therein lies a tale—or rather, conflicting tales. Lorelei is described variously as a "sorceress" (*Stories* 1870, 67), "siren" (*Encyclopedia Britannica* 1960), "water nymph" (Leach 1984, 645), "mermaid," and even, in the plural, "mermaids" (Conway 2002, 164). In any case, at least she represents a romantic legend of the Rhine—or does she?

Introducing Lorelei

My notes on Lorelei remained in my files gathering dust for a decade until I came across a tattered old booklet, *Stories and Legends of the Rhine between Worms and Cologne* (1870), in an antique shop. It was in English and I bought it at once, discovering therein that "Lorelay" [sic] was included. The entry consisted mostly of two poems, a ballad by Clemens Brentano (1772–1842) and a shorter poem by Heinrich Heine (1797–1856). I give the latter, in my own translation, in the accompanying panel.

Folklore or Fakelore?

Now the Rhine has long been a source for romantic tales, and the German epic poem *The Nibelungenlied* (ca. 1200) associates it with a dragon, a treasure of gold, a cloak of invisibility, and other fabulous elements (*Benét's* 1987, 692; Leach 1984, 791).

Yet the Lorelei narrative is, in fact—in folkloristic terms—"neither myth

206

nor local legend" but is rather a "fabrication" by the previously mentioned Clemens Brentano (Leach 1984, 645). Genuine *folklore* consists of traditions (tales, customs, rituals, songs, etc.) accumulated through folk transmission. It includes not only the simple *folktale* but also the *legend* (a localized narrative that is more historicized than the folktale) and the *myth* (which presents preternatural topics as explanations or metaphors of cosmic or natural forces or the like; folklorists do not use the term to mean "a false belief") (Brunvand 1996; *Benét's* 1987).

But whereas folklore is the product of tradition, *fakelore*—a spurious form named by great folklorist Richard M. Dorson (1950)—is deliberately created, as by writers. For instance, many of the tall tales about American herculean logger Paul Bunyan "were literary embellishments of a small amount of oral tradition" produced by William B. Laughead, a lumber company advertising executive (Walls 1996). (Sadly, such distinctions are often confused, as by one pop skeptic who declared that a certain story was "a legend" that he then branded as complete "fiction" by a certain author!)

In his Lorelei fakelore—represented by a ballad (a narrative in verse form) inserted in his novel *Godwi* (written 1800–1801)—Clemens Brentano became "the first to associate the [Lorelei] rock with a woman of the same name." However, "The poem is so convincingly folklike in style that Brentano's invention came to be regarded as a genuine folk legend" (*Benét's* 1987, 581).

Heine's Siren

Indeed, Heinrich Heine may well have regarded Brentano's ballad as presenting a legend, referring to the Lorelei story as *Ein Märchen aus Uralten* (i.e., "an ancient folktale"). Or perhaps he was simply following Brentano's lead in presenting a newly written tale as a handed-down one in order to provide what writers call "verisimilitude" (from the Latin *verisimilis*; *verus*, true, and *similis*, like), that is, a semblance of being true or real.

In any event, whereas Brentano's "Loreley" was a Zauberin ("sorceress"), it remained for Heine (ca. 1823) to create the concept of Lorelei as a siren whose singing lured boat-men to their destruction *(Benét's* 1987,

581).

Of course, Heine did not invent sirens. As far back as the ninth century BCE, the Greek poet Homer in his epic poem *Odyssey* presented sirens: half-woman, half-bird creatures whose singing so enticed sailors that they died by forgetting to eat. To escape their irresistible attraction, Odysseus (Ulysses in Latin) filled his men's ears with wax and had himself lashed to his ship's mast (*Benét's* 1987, 904). Sirens are not to be confused with another woman/bird hybrid, the Harpies. Those were hideous, vulturelike monsters that seized the food of victims and otherwise tormented them (Nickell 2011, 201–202).

Only in some later traditions were sirens depicted as mermaids (Nickell 2011, 201–202). In his great poem, "The Love Song of J. Alfred Prufrock," T.S. Eliot (1915) plays on this tradition when "Prufrock" laments his inconsequentialness:

Shall I part my hair behind? Do I dare to eat a peach?

I shall wear white flannel trousers, and walk upon the beach.

I have heard the mermaids singing, each to each.

I do not think that they will sing to me.

Neither bird-woman nor mermaid, Heinrich Heine's "The Lorelei" was no hybrid but simply a water nymph or beautiful Rhine maiden such as is now represented at the Lorelei/Loreley rock: on the rock itself is a stone sculpture depicting her, and at the base another, in bronze (which appears on picture postcards) (Zieman 2000).

* * *

Ironically, it was by way of Heine's poem that the pseudo-legend of Lorelei finally did become something more than fakelore. The poem attracted English readers, and—especially when set to music by Friedrich Silcher (adapting a folk song [*Encyclopedia Britannica* 1960])—became to tourists "a

208

local legend of sorts." Also because the rock is attended by a peculiar echo, "the romantic literary fiction has had an excuse for passing to a degree into tradition" (Leach 1984, 645).

References

Benét's Reader's Encyclopedia, 3rd ed. 1987. New York: Harper & Row. 2012.

Brunvand, Jan Harold, ed. 1996. *American Folklore: An Encyclopedia*. New York: garland Publishing.

Conway, D.J. 2002. *Magickal Mystical Creatures*. St. Paul, MN: Llewellyn Publications.

Dorson, Richard M. 1950. Folklore and fakelore.

The Lorelei

By Heinrich Heine
(Translated by Joe Nickell)

I know not what it means,
This sadness that I find;
But an olden tale, it seems,
Has overcome my mind.
The air is cool at sunset,
And quietly flows the Rhine;
In the fading evening light,
The mountain summits shine.

The fairest maiden dwells
In marvelous radiance there;
Arrayed in gleaming jewels,
She combs her golden hair.
Combing with a golden comb,
All the while sings she,
A song with a wondersome,
Overpowering melody.

The boatman in his little craft—
Captivated by its might—
Sees not the looming reef,
Stares only at the height.
I think the waves are devouring
Both boat and boatman gone;
And all this with her singing
The Lorelei has done.

Zanzibar's Popobawa: Demonic Monster That Attacks Skeptics

In 1995 I published a short article titled "The Skeptic-raping Demon of Zanzibar," telling of a bat-winged, cyclopean dwarf that reportedly swept into bedrooms and attacked men—especially those who disbelieved in the creature. The phenomenon had occurred in previous decades but had returned. A colleague handed me an article on the phenomenon and joked, "Here's a case for you to solve." Reading a few paragraphs, I replied, "I have solved it."

On the Case

According to *The Guardian* (McGreal 1995), a victim at first thought he was dreaming but felt something pressing down on him—no doubt the Popobawa (the name is Swahili for "bat-wing") who had come to sexually attack him. (See Figure 23.1.) I recognized the phenomenon as having the characteristics of a common "waking dream" (or hypnagogic experience). This occurs when the percipient is in a state between being asleep and awake, and exhibits features of both: A person has a dreamlike (hallucinatory) experience while seemingly awake; the sense of being held down—called "sleep paralysis"—comes from the body's still being in the sleep mode. I traced the phenomenon to many places and times, including the incubus of medieval Europe.

The hypothesis seemed to explain most of the reported Zanzibarian attacks.[1] It follows the principle of Occam's razor, that the explanation making the fewest assumptions is most likely to be correct.

Figure 23.1. Artist's conception of the Zanzibarian Popobawa. (Drawing by Joe Nickell)

My observations began to be cited (or occasionally borrowed without attribution), finally gaining some prominence in a book, *Popobawa* by Katrina Daly Thompson (2017), a professor of African Cultural Studies. She reports my insights on the waking-dream (i.e., hypnagogic) phenomenon and acknowledges that I was "the first to put forth this hypothesis" for the *Popobawa* attacks. But she also seems to resent me for it. She points to a couple of textual simplifications I made in republishing my article and hints at some ulterior motive. Actually, the changes were practical ones in transitioning from a newsletter for fellow skeptics to a book for a general audience (Nickell 2010).

Another example of her accusatory tendency is her finding—wrongly— that I was unfairly "associating Zanzibaris with fear and Westerners with skepticism" (Thompson 2017, 174). To the contrary, I actually gave several examples of Western waking-dream panics.

Field Investigation?

But if Thompson figuratively mussed up my hair, she ran over Benjamin Radford with a truck! Radford, visiting Zanzibar in 2007, took the opportunity to do what he called "the first full field investigation" into the

Popobawa. His article appeared in *Fortean Times* (Radford 2008). However Thompson finds his efforts "fundamentally flawed in content and methodology." She repeats his own admission to having wasted much time at a library, then more seriously accuses him of "misrepresenting secondary sources as if they were primary voices he encountered in the field."

She further insists that his "own information contradicts his claim that Popobawa appears periodically and contemporaneously with 'Muslim holy days' or with election cycles" (Thompson 2017, 165–171). Ironically, this is the very information he had gone in search of: the "cultural context" missing from the waking-dreams explanation (Radford 2008). Her evidence appears to undermine his conclusion that there is a simple pattern to the attacks.

Conclusions

What is Thompson's own view of the Popobawa? She seems to grudgingly accept the psychological explanation of the waking dream, while insisting on the obvious: that it only applies to those cases where the evidence warrants, and that there are also the powerful influences of popular discourse and even jokes (Thompson 2017, 171–177). Indeed, I suggest the list could well include hoaxes, journalistic distortions, elements of mass hysteria, and so on and on. Yes, Westerners should be wary of imposing simplistic patterns on another culture, but they also should not shy away from making scientific observations where appropriate.

Note

1. Radford (2018) wrote a lengthy reply.

References

McGreal, Chris. 1995. "Zanzibar Diary," *The Guardian*, October 2.

Nickell, Joe. 1995. "The Skeptic-raping Demon of Zanzibar," *Skeptical Briefs*, December, p. 7.

———. 2010. *The Mystery Chronicles: More Real-Life X-Files*. Lexington: University Press of Kentucky, 124–127.

Radford, Benjamin. 2008. "Popobawa! In Search of Zanzibar's Bat-Winged Terror." *Fortean Times* No. 241 (November), 34–39.

———. 2018. Popobawa Vs. the Skeptics. Online at centerforinqury.org; accessed November 26, 2018.

Thompson, Katrina Daly. 2017. *Popobawa: Tanzanian Talk, Global Misreadings*. Bloomington: Indiana University Press.

Investigating Werewolves at Moosham Castle

Supposed inhibitors of the night, or at least the nightmares of the credulous, are vampires, zombies, and werewolves—allegedly supernatural man-beasts.

The term werewolf literally means "man-wolf" (from the Old English *wer*, "man," and *wulf*, "wolf") and describes either a human being who has been turned into a wolf by sorcery or one who makes the transformation (whether by will or otherwise) from time to time. In European folk belief, the werewolf preyed on humankind each night but returned to human form at the light of dawn. It could only be killed by being shot with a silver bullet (Leach 1984, II; King 1991, 114).

Becoming a Werewolf

The concept that a human could turn into a wolf seems to have originated with the simple wearing of an animal robe for warmth, with people coming to believe that the man wearing the skin took on the animal's powers. Eventually, the popular imagination conceived of bewitched men who, under the full moon's irresistible power, grew hairy coats, fangs, and claws and otherwise took on the aspect of a beast. The wolf was a popular form of such metamorphoses in Europe.

In fact, there are two medical conditions that undoubtedly helped foster belief in werewolves. One is a disease, a hormonal disorder known as Cushing's Syndrome, which can produce enlargement of the hands and face, together with rapid and copious growth of hair on the latter and an accompanying "acute emotional agitation." According to occult critic Owen Rachleff (1971, 215), "Individuals afflicted with this disease, either because of ostracism or because of the psychotic ramifications of their illness, were, in the past, forced to live apart from society."

There is also the psychiatric disorder known as lycanthropy (after the Greek *lykanthropos*, "wolf-man"). This is the delusion that one has been transformed into a wolf, which can cause sadistic and even cannibalistic or necrophilic behavior (Stein 1988, 37).

The moon is not a factor (except perhaps a psychological one) in cases of "real" werewolves (Rachleff 1971, 215); however, something of the concept nevertheless survives in the popular notion of "moon madness." Also known as the lunar effect, it is the supposed influence the moon exerts on people's behavior. As psychologist Terence Hines explains, when moon-madness proponents' studies are scrutinized, invariably "methodological or statistical flaws have appeared that invalidate the conclusions," and the overall data on the effect "shows overwhelmingly that the moon's phase has no effect on human behavior" (Hines 1988, 157, 158).

It should come as no surprise that lycanthropy is closely associated with vampirism, including a popular belief that one dying under the werewolf's curse was doomed to return as a vampire. In Slavic countries, certain names for werewolves were in time applied to the undead (e.g., *vrykolakas*, *volkodlak*). Also, French demonologists described a type of werewolf, a *loublin*, that haunted cemeteries, digging up and devouring corpses (Bunson 1993, 279–280; Thorne 1999, 72, 91).

Werewolf Craze

Werewolves were part of the witch craze of the sixteenth and seventeenth centuries, particularly in Europe. There, thousands of werewolf cases were reported from 1520 to 1630 (Bunson 1993, 279).

For example, in France in the early 1500s, three men were put on trial for transforming themselves into werewolves and killing sheep. They were convicted and burned at the stake (Rachleff 1971, 216). Near the end of the century in 1598, a French beggar named Jacques Roulet was also tried as a werewolf. Discovered hiding in the bushes near the mutilated body of a teenage boy half-naked and smeared with blood, Roulet admitted to the murder. However, invoking a popular belief of the time, he blamed a magic ointment that he said caused him to become a wolf (whether physically or mentally is unclear). Although he was sentenced to death, on appeal the Parlement of Paris instead committed him to an insane asylum for two years (Stein 1988, 33).

Some have speculated that in such cases belladonna, herbane, aconite root (wolf's bane),[1] or other potent drugs were included in the "witch ointment." One speculator, Dr. H.J. Norman (1966, 291), concluded, "The chief effect was brought about as the result of the high degree of suggestibility of the individuals, who were undoubtedly in numerous instances psychopathic and mentally deranged." No doubt even more important in many cases was the effect of torture, which may have caused the accused "werewolf" to acknowledge the use of whatever the inquisitors imagined—ointment or otherwise.

In one instance—the case of Peter Stump or Stub who was executed near Cologne in 1590—the catalyst was a "girdle" he supposedly put on and took off, thus transforming himself into a wolf and back. Apparently a serial-killer similar to Jeffrey Dahmer, Stump raped, murdered, and even devoured men, women, and children. His was "one of the most famous of all German werewolf trials" (Summers 1966, 253). Revealingly, when his interrogators could not find the magical girdle where the confessed lycanthrope said he had discarded it, they "supposed that it was gone to the devil from whence it came" (quoted in Summers 1966, 259).

Figure 24.1. Werewolf attacks a man (Wood engraving published in Strasburg in 1561.)

Torture at Moosham

While on an investigative tour of Europe in May 2007, I came across a much later werewolf case in Austria. German skeptic Martin Mahner and I toured a supposedly haunted Schloss Moosham (i.e., Moosham Castle [Figure 24.2]) where many witch trials were held. Between 1675 and 1689, when the witch mania had already decreased elsewhere, some 200 victims were executed, mostly vagabonds.

The werewolf scare occurred still later, between 1715 and 1717, when an unusual number of cattle and deer were killed by wolves in the Moosham district. When attempts to hunt down and kill the predators failed, superstitious folk concluded that the creatures must have been supernatural. Subsequently, two adolescent beggars admitted under torture in the Schloss Moosham dungeon—where I have been—to receiving a black cream from the Devil. Had they put the unguent on their bodies, they confessed, they would have been transformed immediately into wolves. The implication was that people conspired with the Devil to turn into wolves and were responsible for the animal killings. Needless to say, neither the existence of the alleged ointment nor its effect was ever demonstrated.

Figure 24.2. Moosham Castle in Austria, where witches were tried and where in 1717 men were tortured into confessing involvement in werewolf attacks. (Photo by Joe Nickell)

In this instance, the Devil's confessed accomplices escaped execution. They were instead reportedly sentenced to lifelong service as Venetian galley slaves, a punishment described as "a slow but sure death" (Bieberger et al. 2004, 157–162).

Further evidence of the Moosham Castle werewolf case turned up (as director of CFI Libraries Timothy Binga discovered while searching online sources) in an archive of werewolf reports from 1407 to 1720 (Werwolfprozesse 2002). There are two listings for the year 1717 in Moosham: the first, Philipp Ebmer, a beggar, was noted as having died in detention; the second was Ruepp Gell, who, with Hans Pfaendel and five other codefendants, all beggars, ultimately "died in detention" after being sentenced to *Galeerenstrafe*, or "galley-punishment" (Werwolfprozesse 2002).

As a replacement for the death penalty, during *Galeerenstrafe* the condemned man was secured with heavy iron chains to a galley's rudder. This inhumane punishment typically resulted in death by exhaustion, disease, or shipwreck (Galeerenstrafe 2007).

We like to ascribe such frightening excesses to the magical thinking that pervaded an earlier age, holding our own time as more enlightened. Yet we must acknowledge the surprisingly modern view of lycanthropy found in the sixteenth-century skeptical work, *The Discoverie of Witchcraft*, by Reginald Scot (1584, 58). Challenging the basis of claims that men can be transformed into beasts, Scot sums up:

> To conclude, I saie that the transformations, which these witchmongers doo so rave and rage upon, is (as all the learned sort of physicians affirme) a disease proceeding partlie from melancholie, wherebie manie suppose themselves to be woolves, or such ravening beasts. For *Lycanthropia* is of the ancient physicians called *Lupina melancholia*, or *Lupina insania*. J. Wierus declareth verie learnedlie, the cause, the circumstance, and the cure of this disease. I have written the more herein; bicause hereby great princes and potentates, as well as poore women and innocents, have beene defamed and accounted among the number of witches.

Conversely, we must also acknowledge some of the unenlightened thinking of today. Consider, for example, the "animal mutilation" cases that burgeoned in the 1970s and continue to the present. They are often attributed to the "chupacabra," an imagined Draculaesque extraterrestrial, despite repeated evidence that the "mutilations" are the work of predators and scavengers (Nickell 2006, 20–21). Perhaps some of us have not advanced so very far after all.

Notes

1. Aconite, or wolf's bane, is a very poisonous plant, often "added to protection sachets, especially to guard against vampires and werewolves" (Cunningham 2000, 260). It was also placed before windows and doors (Bunson 1993, 283).

References

Bieberger, Christof, et al. 2004. *Geisterschösser in Osterreich* ("Ghost Castles in Austria"). Vienna: Verlag Carl Ueberreuter. (Portions translated for me by Martin Mahner.)

Bunson, Matthew. 1993. *The Vampire Encyclopedia*. New York: Gramercy Books

Cunningham, Scott. 2000. *Cunningham's Encyclopedia of Magic Herbs*, 2nd ed. St. Paul, Minnesota: Llewellyn Publications.

Galeerenstrafe. 2007. From German wikipedia.org; accessed July 18, 2007.

Hines, Terence. 1988. *Pseudoscience and the Paranormal*. Amherst, N.Y.: Prometheus Books.

King, Francis X. 1991. *Mind & Magic*. London: Crescent.

Leach, Maria, ed. 1984. *Funk & Wagnall's Standard Dictionary of Folklore, Mythology, and Legend*. New York: Harper & Row.

Nickell, Joe. 2006. Argentina mysteries. *Skeptical Inquirer* 30(2) (March/April), 19–21.

Norman, H.J. 1966. Witch ointments. Appendix to Summers 1966.

Rachleff, Owen. 1971. *The Occult Conceit*. Chicago: Cowles.

Scot, Reginald. 1584. *The Discoverie of Witchcraft*. Reprinted (from a 1930 ed.) New York: Dover, 1972, 58.

Stein, Gordon. 1988. Werewolves. *Fate* magazine, January, 30–40.

Summers, Montague. 1966. The Werewolf. New York: Bell Publ. Co.

Thorne, Tony. 1999. *Children of the Night: Of Vampires and Vampirism*. London: Victor Gollancz.

Werwolfprozesse in der Frühen Neuzeit. 2002. Available online at accessed July 13, 2007.

At a Vampire's Grave

Given the ubiquitousness of vampires, those undead beings who are driven by bloodlust (and who thrive in movies like 2008's popular *Twilight*), it should not be surprising that historically there have been instances of reputed vampirism in the United States, notably in New England. I have investigated these cultural trends on site, even tracking the legendary creatures to their very graves (Nickell 2009).

New England has always been an admixture of both austere skepticism and passionate superstition. Vampire legends lurk in the latter. According to one vampirologist, "The presence in New England of a strongly rooted vampire mythology is something of an enigma to folklorists. There is quite simply no other area in all of North America with such wealth of vampire lore" (Rondina 2008, 165).

Mercy Lena Brown

One of the best known examples is the case of nineteen-year-old Mercy Lena Brown in Exeter, Rhode Island, in 1892—a case that supposedly influenced Bram Stoker, author of *Dracula* (1897). As Katherine Ramsland (2002, 18) concisely tells the story:

> George Brown lost his wife and then his eldest daughter. One of his sons, Edwin, returned and once again became ill, so George exhumed the bodies of his wife and daughters. The wife and first daughter had decomposed, but Mercy's body—buried for three months—was fresh and turned sideways in the coffin, and blood dripped from her mouth. They cut out her heart, burned it, and dissolved the ashes in a medicine for Edwin to drink. However, he also died, and Mercy Brown became known as Exeter's vampire.

> Accounts of the exhumation in the *Providence Journal* of March 19 and

21, 1892, acknowledge that the Browns died of consumption (tuberculosis). They do not mention the corpse of Lena (as she was actually known) being turned on its side or blood dripping from the mouth. The exhumation was conducted by a young Harold Metcalf, MD, from the city of Wickford. "Dr. Metcalf reports the body in a state of natural decomposition, with nothing exceptional existing," stated the *Journal*. "When the doctor removed the heart and the liver from the body a quantity of blood dripped therefrom, but this he said was just what might be expected from a similar examination of almost any person after the same length of time from disease." The article added, "The heart and liver were cremated by the attendants" ("Exhumed" 1892).

A follow-up article ("Vampire" 1892) noted that the heart's blood was "clotted and decomposed . . . just what might be expected at that stage of decomposition." The correspondent acknowledged the custom of an afflicted person consuming the ashes to effect a cure, stating, "In this case the doctor does not know if this latter remedy was resorted to or not, and he only knows from hearsay how ill the son Edwin is, never having been called to attend him."

A Post-mortem on Case

And so ends "Unarguably the best known incident of historical vampirism in America," indeed the story of "The Last Vampire" (Rondina 2008, 83, 99). However, there are many other reported cases typically involving consumption. The victim's lethargy, pale appearance, coughing of blood, and contagiousness all suggested to the superstitious the result of a "vampire's parasitic kiss" (Citro 1994, 71).

In 2016 I visited the grave of the much-abused Lena (Figures 25.1 and 25.2). (I had been in Rhode Island at the invitation of Norma Sutcliffe who owns the eighteenth-century property central to the claims behind the horror movie *The Conjuring*, 2013.) Returning home, I detoured to Exeter where Lena lies, an iron strap preventing her tombstone from being stolen. Coins and other small items are placed thereon, either as gifts to spirits, or with some other intention, often some thought of warding off evil. When we look back at superstitious eras, we should not think our generation too superior.

Figure 25.1. Grave of the "Exeter Vampire" (Exeter, RI): Mercy Lena Brown.

Figure 25.2 Vault in cemetery where Brown's coffin was reportedly held for reinterment. (Photos by Joe Nickell)

References

Citro, Joseph A. 1994. *Green Mountain Ghosts, Ghouls and Unsolved Mysteries*. Boston: Houghton Mifflin.

"Exhumed the Bodies. . . ." 1892. *Provincetown Journal*, March 19 (reprinted in Rondina 2008, 86–87).

Nickell, Joe. 2009. Searching for Vampire Graves. *Skeptical Inquirer* 33: 1 (March/April), 16–19.

———. 2018. Visiting the grave of the last vampire. Investigative Briefs blog. Online at https://centerforinquiry.org/blog/visiting_the_grave_of_the_last_vampire/; accessed August 29, 2018.

Ramsland, Katherine. 2002. *The Science of Vampires*. New York: Berkley Boulevard Books.

Rondina, Christopher. 2008. *Vampires of New England*. N.p.: On the Cape Publications.

"The Vampire Theory." 1892. *Providence Journal*, March 21; reprinted in Rondina 2008, 89–96.

In Search of the 'Chupacabra'

Over my years as a skeptical cryptozoologist—writing numerous articles and books about my investigations of strange creatures, as well as appearing on such television shows as National Geographic's *Is It Real?* and the History Channel's *MonsterQuest*—I have taken a simple approach. I have found that it is often possible to identify a real, natural world lookalike for a given fabulous creature, one that could be mistaken for it.

While a naturalist is one who studies the natural world using the methodology of science, I use the term *paranatural naturalist* to describe one who applies the naturalist's methods to mysteries of the paranatural with the intention of solving them. Here is an example of that approach applied to a relatively modern monster.

The Chupacabra Appears

This fabled vampiric creature has reportedly preyed on farm animals in many countries since it first appeared in Puerto Rico in 1995. The "original" sighting was in Puerto Rico about the second week of August around 4:00 p.m. A housewife was looking out the window when a creature—walking on two legs—came and stood nearby for about three minutes. The woman judged the creature to be four feet tall but later reduced that to three. Its chest was thin, but it was more "chubby" below with distinct "hips." It had a "flat" face, but her view was a direct frontal one. It had "protruding" black eyes, two holes for a nose, and a straight "slash" of a mouth.

Most notably, the creature's thin upper limbs were drawn back as if from injury. That appears to account for its walking—and even hopping "like a kangaroo"—on its hind legs. Indeed, there were apparent burns on its body, which the woman's husband, having seen it in the morning in his machine shop, thought were from battery acid. The animal's general morphology points to one of the numerous wild dogs of the island. Subsequently, about ten

231

others witnessed the creature, some observing "fangs" (no doubt canine teeth) and "spines"—i.e., hackles: bristles that run along a dog's neck and back and can become erect when the animal shows hostility. These features are corroborative that the animal was from the family *Canidae*. The first witness stated that she looked for—and did not see—genitals, from which I deduce that the crippled dog was a female.

This case provides a good example of how unusual features and bizarre locomotion probably distracted the eyewitnesses from recognizing the animal's true nature. However, as with many cryptids, specific features can often help the investigator make a probable identification. (For more, see Scott Corrales, *Chupacabras and Other Mysteries*, 1997).

I wrote one of the early skeptical pieces on the "goatsucker" (as its Spanish name translates to) after the scare spread to Mexico in April 1996. I was assisted by colleagues in Mexico, Patricia and Mario Mendez-Acosta. As it happened, when authorities staked out farmyards, wild dogs were caught each time. Although the goats, turkeys, horses, and other animals killed by the chupacabra were invariably claimed to have been drained of blood, the reverse was true; necropsies consistently showed that gravity had simply caused the blood to settle (drain downward) in the carcasses.

By 1996, the myth had also reached Florida—as elsewhere, being initially spread by the Spanish-speaking media. The further migration came as no surprise. Involving major livestock areas, the reported mutilations sparked conspiracy theories by UFOlogists, journalists, and local workers.

In the United States, "chupamania" found a ready breeding ground. Animal mutilation claims had been rife in the 1970s, although they were carefully investigated and attributed to the work of predators and scavengers (Frazier 1980; Nickell 1995, 115).

From initial media reports the craze quickly spread via electronic telecommunications. It became the first such monster craze to be diffused by the Internet. According to one media observer, it once took centuries for a monster legend to filter down through generations; now the process is similar except for the increased speed of dissemination (Trull n.d.). Soon the stories

232

were featured in magazine articles, segments of books, and more. As early as 1997, a Hispanic-American cryptozoologist, Scott Corrales, had produced a book, *Chupacabras and Other Mysteries*.

Evolving Myth

Soon, artistic renderings of the chupacabra evolved into a vaguely humanesque entity—at least in walking upright on two legs and having "long, thin arms and hands with three long, skinny fingers with claws" (Coleman and Huyghe 1999, 80). Its large head, wraparound eyes and minimal other facial features (merely two nasal holes and a slash for a mouth) evoked the already popular concept of the gray alien of alleged extraterrestrial encounters (Coleman and Huyghe 1990, 80–81; Coleman and Clark 1999, 61–63).

American chupacabra lore took a new turn when three of the supposed monsters—or at least "three hairless doglike creatures"—became roadkill near Cuero, DeWitt County, Texas, in 2007. When a fourth was discovered in 2008, DNA tests showed the creature was *Canis latrans*, the common coyote. However, a second round of tests at the University of California, Davis—with special expertise in animal forensic science—revealed that it was a cross between a coyote and a Mexican wolf. As to the lack of hair, Loren Coleman has observed that many of these may be dogs or other animals suffering from mange (Nickell 2011a, 144).

My Encounter

In 2011, an animal some speculated was a chupacabra made repeated attacks on the farm of Tim Stoll near Strafford, Missouri, costing the family a chicken a day. On November 5, Stoll's teenage stepson, Dalton Pennington, encountered the creature. It had spooked their horses and was heading for the goat pen, when Dalton killed it with a single shot from a deer rifle. He was impressed with its strange appearance and distinctive yellow eyes, and thought it might be the fabled creature.

Being in Missouri at the time (lecturing at SkeptiCon IV in Springfield). I was able to investigate the case firsthand. I contacted Francis Skalicky, a media specialist for the Missouri Department of Conservation. He told me that two state biologists had—independently—identified the Strafford creature from photos as a coyote with mange.

I also went on site at the Stoll farm where I was very hospitably received. Young Dalton even recognized me from one of my several appearances on *MonsterQuest*. On orders from conservation officials, they were about to dispose of the animal by burning, so I arrived, with Missouri skeptic Larry Jewell who videotaped our visit, just in time to inspect the very rank-smelling carcass.

The creature is indeed weird looking. However, its size, lithe body, doglike paws, and triangular ears are indicative of the canids, and the sharp snout, specific coloring (orangish-gray upper, buff underparts, rusty legs with vertical dark line on lower foreleg), and other features, including the dentition and yellow eyes, are consistent with the coyote (Figures 26.1, 26.2, and 26.3). By no means hairless, the animal nevertheless clearly suffered from mange (Nickell 2011b)—a condition I was familiar with as a boy growing up in eastern Kentucky. (See Whitaker 1996, 682–86.)

Figure 26.1. Coyote with "Chupacabra" disguise (drawing by Joe Nickell, with apologies to Warner Bros. Road Runner).

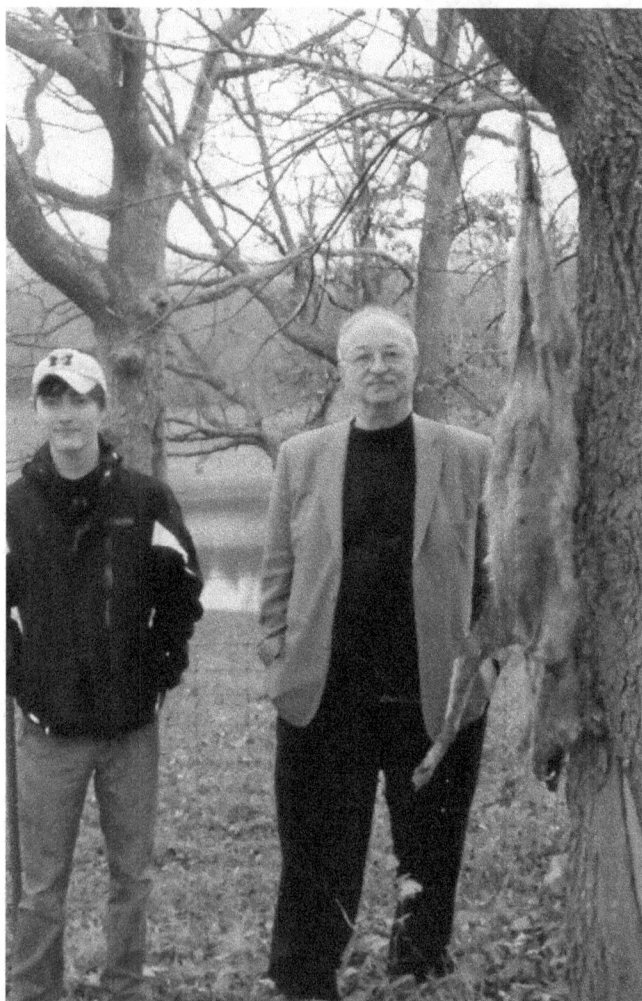

Figure 26.2. Dramatis personae (left to right) Dalton Pennington, Joe Nickell, mangy-coyote "chupacabra." (Author's photo.)

Figure 26.3. Close up of "chupacabra" showing canid ears and dentition. (Author's photo)

In marked contrast, the chupacabra is said to be "hairy, about four feet tall, with a large, round head, a lipless mouth, sharp fangs, and huge, lidless red eyes." Moreover, it reportedly has "thin clawed, seemingly webbed arms with muscular hind legs" and "a series of pointy spikes running from the top of its head down its backbone" (Coleman and Clark, 1999, 61–63). However, since no specimen has ever been authenticated scientifically, the Strafford creature is as real, apparently, as the Chupacabra gets.

References

Coleman, Loren, and Jerome Clark. 1999. *Cryptozoology A to Z*.
New York: Fireside.

Coleman, Loren, and Patrick Huyghe. 1999. *The Field Guide to Bigfoot, Yeti, and Other Mystery Primates Worldwide*. New York: Avon Books.

Corrales, Scott. 1997. *Chupacabras and Other Mysteries*. Murfreesboro, TN: Greenleaf.

Frazier, Kendrick. 1980. "Cattle Mutilations: Mystery Deflated, Mutologists Miffed." *Skeptical Inquirer* 5, no. 1 (Fall): 2–6.

Los Angeles Times. 1996. May 19.

Nickell, Joe. 1995. *Entities*. Amherst, NY: Prometheus Books.

———. 1996. "Goatsucker Hysteria." *Skeptical Inquirer* 20, no. 5 (September/October): 12.

———. 2004. *The Mystery Chronicles*. Lexington, KY: University Press of Kentucky, pp. 28–30.

———. 2011a. *Tracking the Man-Beasts*. Amherst, NY: Prometheus Books.

———. 2011b. "Chupacabra" attack. Online at http://www.centerforinquiry.net/blogs/entry/chupacabra _attack/; accessed November 29, 2011.

Trull, Donald. N.d. Cited in Coleman and Clark, 1999, p. 62 (above).

Whitaker, John O. 1996. *National Audubon Society Field Guide to North American Mammals*. New York: Alfred A. Knopf.

Identifying the Enigmatic 'Dover Demon'

According to *The Field Guide to North American Monsters* (Blackman 1998, 106), "Few cases in the history of cryptozoology have received as much attention as the peculiar Dover Demon affair" of 1977. All these years later, that remains even more true. But was the subject itself true? Was it really some hitherto unknown and seemingly impossible creature? Or a real animal seen by four badly mistaken witnesses? Or even, as local police first thought, "nothing more than a school vacation hoax" (Associated Press 1977)? One more question: Shall we really dare to take on such a case–especially after so much time has elapsed?[1]

Three Encounters, Four Witnesses

The strange affair began about 10:30 p.m. on the night of April 21, 1977, when Bill Bartlett (age seventeen) caught a strange-looking creature in his Volkswagen's headlights while driving with two friends along a dark road in Dover, Massachusetts. (The town, then with a population of about 5,000, consisted of mostly woods and pastures, some fifteen miles southwest of Boston.) Bartlett claimed he had seen the creature beside a low rock wall on his left and that he first thought it was a cat or dog until he saw it had an extremely large head. It turned to look at him with shining orange eyes as he drove by.

Bartlett asked his friends if they had seen the being. He appeared visibly shaken. They told him they had not been looking in that direction and urged him to turn around and drive back. However, they saw no sign of the monster. Bartlett described it as some three-and-a-half-feet tall, having a watermelon-shaped head with big eyes (albeit lacking any other features), and a very thin body with "spindly" arms and legs. He claimed it was a peach color

240

but was "almost whitish" near its very long fingers (Webb 1977, 42–47). Subsequently, he made at least two drawings of what he reported having seen (Webb 1977, 43–44).

A second encounter about two hours later was reported by John Baxter, age fifteen, who was walking home from his girlfriend's house. About 1.2 miles northeast of the Bartlett location, Baxter saw the silhouette of a "very small" figure walking toward him. He thought it might be a kid he knew and called out, but the figure did not respond. When it was about fifteen feet away (approximately twenty-five feet by later investigators' measurement), it stopped, and Baxter too halted. Suddenly the entity scurried away down an embankment, across a shallow wooded gully, and up the opposite side. Baxter started in pursuit, but after a few steps he stopped and, feeling uneasy, backed away (Webb 1977, 46–48).[2]

Baxter said he could hear the small creature running in the dry leaves; it came into view again across the gully (about thirty feet away by the investigators' later tape measurement). Baxter said he thought the dark silhouette resembled a monkey's, except that it had a large "figure-eight" head (as distinguished from the shape of a watermelon, which does not have double lobes). He told investigators its feet were "molded" around a rock near a tree, and its lengthy fingers appeared to grip the trunk—as for support. He could not say how many digits were on its hands or feet. Baxter thought the creature appeared as if it might be ready to spring at him. So he "walked very fast" back to the street and on to an intersection, where he caught a ride home. He thought his encounter lasted less than ten minutes—"probably considerably less"—concluded the report (Webb 1977, 47–49).

The third encounter took place the following evening (April 22), nearly twenty-four hours after John Baxter's sighting. There were two witnesses, Will Taintor (age eighteen) and his girlfriend, Abby Brabham (age fifteen). Taintor and Brabham glimpsed the "tan" creature, resembling a monkey, crouched in the road. Apparently neither of the two noticed hands or feet, but they described the entity as having a large "oblong" head (again, not double-lobed) and eyes that shone "bright green" (she insisted) in the headlights. Taintor then sped up at Brabham's urging. Investigators' subsequent timed reenactment placed the duration of the pair's observation at approximately five seconds (Webb 1977, 49–50).

Account Assessment

Attempts to identify a real animal behind the sightings have not fared well. For example, one theory was that it was a baby moose (Kottmeyer 1998). However, its large size, four-legged stance, and big ears render that idea most unlikely.

Nevertheless, three encounters by four individuals have been taken as corroboration of what has been described, investigators having been at pains to characterize the eyewitnesses as truthful. Still, there are some problems.

First, Bill Bartlett would later admit he had smoked marijuana approximately an hour before his sighting, although he believed he was clear-headed "during the crucial moments of the encounter" (Webb 1977, 45). Also, the fact that John Baxter wrote sci-fi stories was, according to the investigators, "a troublesome feature that can't be completely ignored" (Webb 1977, 52).

Then there is the issue of collaboration. The witnesses are frequently treated as if they had completely independent observations; however, that seems not to be the case. When Baxter (the second to see the creature) made a drawing of what he saw, it was purportedly before he had learned of Bartlett's sighting and drawing, according to Webb (1977, 48), who actually uses the word *allegedly*. There indeed seems reason to doubt Baxter's complete lack of foreknowledge—especially given the striking similarity of very minor details. Such agreement includes the feet being "molded" to rocks and the apparently changed shape of the head: from Bartlett's original "watermelon" and "oblong" to both showing a figure-eight drawing (i.e., peanut-shaped). In other words, I think Bartlett may have been influenced soon by Baxter rather than entirely the other way around (Webb 1977, 46, 48).

Certainly Abby Brabham's boyfriend, Will Taintor, had learned about

242

Bartlett's original sighting—before Taintor and Brabham had their encounter. Taintor told investigators he asked Brabham leading questions, supposedly as a means of testing her. Then, later, he divulged to her what he knew. (Her insistence that the eyeshine was green rather than orange is supposed to bolster her independence. However, it is difficult not to notice how both Brabham and Bartlett apparently made reference to a monkey-like body; Webb 1977, 48–49.) Again, Brabham said the creature's head was "oblong," consistent with Bartlett's original "watermelon-shaped," but emphatically at variance with later sketches by Bartlett and Baxter showing enormous double-lobed heads.

Getting Real

These problems remaining, I believe we can nevertheless still quite credibly place an identified suspect at Dover, Massachusetts, in April 1977—a bipedal creature with a large, round, earless head; missing nose and mouth; and having shining eyes; one standing on spindly legs with ... well, read on.

The creature I have in mind is most active between early evening and early morning. Belonging to the order *strigiformes*, it is among the easiest of its family to identify, yet it is one of the world's least understood. Although it was given its own genus, *Nyctea*, in 1809, in 2003 it was reclassified into the *Bubo* genus (Bannick 2020, 7–52).

We are identifying *Bubo scandiacus*. It is interesting that its closest living relative is *Bubo virginianus*, because that creature actually piqued the interest of the cryptozoologist who gave the Dover Demon its popular name! He is Loren Coleman (whom I know personally). Of him, Walter Webb (1977, 55) states:

Loren was the first to note some resemblance between the Dover entity and the famous Kelly, Kentucky, "little men" incident of 1955. Yet major differences abound, not the least of which were the Kentucky creatures' large floppy ears. Like the Dover Demon, the Kelly being has never been reported before or since.

Now that is no longer true. I identified the Kelly beings as a pair of the genus *Bubo virginianus* (Nickell 2011, 167–173).

In other words, the Kelly beings were great horned owls and the Dover Demon, we can now spell out, was the snowy owl—indeed itself a subject of myth and mystery, including its anthropomorphized version as Harry Potter's pet, Hedwig. Snowy owl experts, including wildlife photographer Paul Bannick (2020), have done much to teach us the secret truths about this fascinating creature.

Snowy owls breed on the Arctic tundra but exhibit a broad range of movements—some flying hundreds of miles for a variety of complex reasons. They tend to winter as far south as the northern United States but will rarely be seen in a tree. They prefer open country; one may sit on a rooftop, post, or other resting place such as rocky ground. It also prefers to be near water (Bannick 2020, 58; Alsop 2001, 372; Bull and Farrand 1994, 548; "Snowy owl" 2023).

Identification

Let us now take a new look at the Dover Demon, recognizing its true identity behind the longstanding camouflage of misperception. It is a large, round-headed owl, indeed one of the world's biggest. It is the only white owl in North America—keeping in mind that the females and juveniles have a relatively large amount of dusky barred (or dash-like) markings, in contrast to the older males, which are effectively all-white (Bull and Farrand 1994; Bannick 2020, 27). The two genders are about the same total body length of approximately twenty-four inches, but the female is roughly 25 percent heavier than the male (Bannick 2020, 27). For all these reasons, we can be confident that the Dover Demon was a male with its more svelte look.

Clearly the purported height of the Dover Demon is excessive for it to have been a snowy owl, which would have been unlikely to be more than about twenty-seven inches tall (Alsop 2001, 372). However, the height of unknown creatures is often exaggerated upward as we see in the Kelly, Kentucky, case. There, for instance, the height of the creatures was first estimated at four feet, but a main witness later downsized that to "two and a half feet tall" (Nickell 2011, 171)—more compatible to the height of a great horned owl at about twenty-five inches (Alsop 2001, 371). Had the Dover Demon's estimated height of three-and-a-half to four feet (given by Bartlett [Webb 1977, 46]) been reconsidered by investigators, I suspect it too would have been downsized. (Still, a snowy owl that had surprisingly appeared in southern California toward the end of 2022 was, according to one eyewitness, "absolutely ginormous," although of course no actual measurement was available [Levenson 2023].)

The snowy owl is often called "earless" (like the Dover Demon) because, instead of external ears, it has only openings on either side of its head hidden by tiny "ear tufts." (These may become noticeable when the owl flies into a strong wind [Bannick 1977, 27, 32, 39].) Similarly, the Dover Demon's lack of a nose or mouth—making the head appear featureless except for the eyes (Webb 1977, 32)—was due to the owl's small beak being nearly hidden by surrounding long, white, dense bristles (Bannick 1977, 32; "Snowy owl" 2023). This apparent absence of ears, nose, and mouth represents a powerful set of similarities between the Dover Demon and the real creature behind it, the snowy owl.

A laughable error in the Bartlett and Baxter drawings—so *mutually* wrong and thus probably again indicative of cross-pollination—is the absurdly thin neck that would have been "anatomically impossible" (Radford 2023, 32). Now owls are generally spoken of as having no neck, as if the head were placed directly atop the body (see Bull and Ferrand 1994, 297–310). However, Bartlett and Baxter each drew the Dover Demon with an enormous double-lobed head supported by a thin neck. The similar error in each drawing bespeaks of one person misleading the other on the issue. Indeed, a possibility for the double-lobed-head look is simply the natural narrowing between the head and upper body—the actual neck area (see Alsop 2001, 372). Remember, the Dover Demon witnesses did not really know what they were looking at. Thus the result of this narrowing is to create—by combining head and shoulders—the "figure eight" feature.

Another illusion-caused error involves the "arms" with "hands." As shown by the Kelly, Kentucky, creatures (the great horned owls), when these owls are not seen in flight—and not understood to be owls—their partially opened wings may be mistaken for "long spindly arms" (Webb 1977, 46), because what is seen then are the wings' bright front edges. (See Bannick 2020, 31, 64, 92, 99.) At the ends of the Dover Demon's "arms" were what both Bartlett and Baxter spoke of as "large hands" or again "long fingers" (Webb 1977, 46, 48). These were the five (or so) splayed feathers at the snowy owl's wingtips. (See examples in Bannick 2020, 34, 92–99. See also Nickell 2011, 167–173.) In fact, the portion of a bird's wing that corresponds to the human hand is called the *manus* (Latin for "hand"). (See Alsop 2001, 731.)

Finally, we need a word on coloration. The male snowy owl is effectively all-white, so its color is affected by light. The Dover Demon was seen in the dark and in the *yellow headlights* of the old cars. Thus, the color of that light was imparted to the feathers, rather than being the actual color of the creature Bartlett said was "peach" (i.e., the color of the yellow-to-light-orange flesh of the peach tree's fruit).

All in all, the Dover Demon proves to have been a quite terrestrial and well-documented creature, recognized at last as the somewhat exotic snowy owl. As a skeptical *monsterologist* of some fifty years' experience, I believe my findings will illustrate to any objective reader that "Snowy" and the "Demon" are truly birds of a feather.

Notes

1. I actually tentatively identified the Dover Demon as a snowy owl about 2005, approximately the time I was extensively studying the earlier Kelly, Kentucky, creatures of 1955 (see Nickell 2011, 167–173). Alas I never completed that study— misjudging the staying power of the Dover Demon tale, which has waxed and waned over the years.

2. Let me go out on a limb (sorry!) and suggest one further clue: a bit of the Dover Demon's behavior toward the eyewitnesses, especially toward John Baxter. Recall that Baxter especially (but also the others to a lesser degree) seems to have been drawn toward the creature by its own actions. Recall, too, that in no instance did it simply take flight to remove itself from a potential pursuer. Instead, it kept on the ground as if attempting to draw the person toward it. Baxter said he was uneasy about the behavior, fearing the being might be attempting to lure him, looking "as if it might be ready to spring" (Webb 1977, 48).

Strikingly, the snowy owl is known to exhibit a particular form of behavior. This is called a "crippled bird act," intended to lure a potential predator away from the owl's nest (Alsop 2001, 372). A nest would be unlikely in the Dover case, but still the act could conceivably have been practiced. (I was once given such treatment by a lakeshore bird called a Killdeer, which feigned a broken wing and, successfully, drew me to follow it for a distance of several feet, then, suddenly was "healed" and flew away! I still regard it as one of the most intriguing things I have seen in nature [Alsop 2001, 233].)

References

Alsop, Fred J. III. 2001. *Smithsonian Handbooks: Birds of North America, Eastern Region*. New York, NY: DK Publishing.

Associated Press. 1977. Teeners report 'creature.' *Bangor Daily News* (May 16): 5. Quoted in Radford 2023, 32.

Bannick, Paul. 2020. *Snowy Owl: A Visual History*. Seattle, WA: Mountaineers Books.

Blackman, W. Haden. 1998. *The Field Guide to North American Monsters*. New York, NY: Three Rivers Press.

Bull, John, and John Farrand Jr. 1994. *National Audubon Society Field Guide to North American Birds: Eastern Region*. New York, NY: Alfred A. Knopf.

Coleman, Loren. 2007. *Mysterious America.* New York, NY: Paraview Pocket Books.

Kottmeyer, Martin S. 1998. Demon moose. *The Anomalist* 6: 104–110.

Levenson, Michael. 2023. Snowy owl awes crowds near L.A. *New York Times* (January 1).

Nickell, Joe. 2011. *Tracking the Man-Beasts: Sasquatch, Vampires, Zombies, and More.* Amherst, NY: Prometheus Books.

Radford, Benjamin. 2023. Deconstructing the Dover Demon. SKEPTICAL INQUIRER 47(1) (May/June): 31–33.

Snowy owl. 2023. Wikipedia. Online at
https://en.wikipedia.org/wiki/Snowy_owl; accessed
January 15, 2023.

Webb, Walter. 1977. Report on the Dover Demon (investigated
with Loren Coleman, Joseph Nyman, and Ed Fogg)
completed September 1977. Given in Coleman 2007,
42–56.

Revealing the Real Carolina Lizard Man

As part of my career as "The Detective of the Impossible," I have investigated the world's "cryptids" (hidden creatures) for well over half a century. I have sought their purported lairs from A to Z—literally from the Adirondack Mountains to remote Zhoukoudian, China. My successes began with an intriguing "devil baby mummy" I encountered in a Toronto curio shop in 1971 and include the "alien" Flatwoods Monster case that had remained from yesteryear but that I solved effectively decades later (Nickell 2011, 148, 159–166). More recently, I identified the very real creature behind the cryptically reported "Dover Demon" of 1977 (Nickell 2023).

Here, I respond to a case from 1988, encountered by a definitive witness (who died in 2009). Despite his apparent misperceptions and erroneous memories, the event nevertheless rings true. Indeed, I am familiar with such a creature I myself have experienced in the wild; I am convinced it is the selfsame one—now no longer a cryptid but an actual resident of swamps and woodlands (Nickell 2011, 177, 179, 180, 215).

By Any Other Name

The story begins in the early-morning hours of June 29, 1988.[1] As teenager Chris Davis drove home from his night-shift job, he experienced a flat tire near Bishopville, a small South Carolina town. As he finished replacing the tire, he heard a noise that drew his attention across a butterbean field of the Scape Ore Swamp. He saw a bipedal figure that had come out of some trees into the light of the almost-full moon. He guessed it was over seven feet tall, assuming it was at a distance of some twenty-five yards, but he gave different estimates in subsequent accounts.

In my experience, the height of unknown creatures is frequently exaggerated by witnesses. For example, something frightening—as in this

250

case—may loom large in one's consciousness (Nickell 2011, 179). Also, distortions of various kinds may occur, especially in cases like that of the Dover Demon (Nickell 2023) in which a witness admits to having smoked marijuana. Moreover, people's memories tend to evolve over time.

In any event, young Davis's attention was caught by the creature's "red eyes glowing." He focused on these as the cryptid rushed toward him. He ran to the car, and locked himself inside. At this point, he could only see the creature "from the neck down" but observed that it had "three big fingers, long black nails and green rough skin" (quoted in Manley 2007).[2] At first he stated the creature had "scales" but later changed that to being "caked with mud" (Bingham and Riccio 1991, 171–174). These descriptors are fairly accurate for the real creature I have identified, as we shall soon see.

Davis continued: "I looked in my mirror and saw a blur of green running. I could see his eyes and then he jumped on the roof of my car. I thought I heard a grunt and then I could see his fingers through the rear windshield, where they curled around on the roof. I sped up and swerved to shake the creature off" (quoted in Manley 2007).

Rushing home, he and his father inspected the vehicle for damage, discovering that the side mirror was twisted and there were deep scratches on the roof.

The report of the vehicle damage has been questioned, seemingly at best exaggerated. Moreover (as Radford [2023, 61] speculates), because Davis "was swerving back and forth through a swampy, wooded area" trying to escape the creature, "it's possible that he scraped low-hanging tree branches." As to the monster's reputed ability to traverse distances with great speed, it is here that we should pause to consider its identity.

Identification

What Chris Davis saw was indeed a creature of the night: a large, stocky, "earless" member of the family Tytonidae—namely, Strix varia, the barred owl. It gets its name from bars (dashes) and other markings on its plumage—i.e., is "heavily streaked, spotted, and variegated brown, buff, and white" (Alsop 2001, 378; Rogers 2008, 162–165). Sometimes they can look like scales or mud caked as Davis indeed stated (Bingham and Riccio 1991,

172).

This owl would be right at home where it was seen because it inhabits "swampy forests." There, in its "dark retreat, usually a thick grove of trees," it rests during the day, coming out in the night to feed (Bull and Ferrand 1994, 551). Of course, its eyes do not "glow," but it indeed has crimson eyeshine, as Davis saw, and is, in fact, the same creature I identified as the culprit in the 1996 flap in West Virginia that gave "Mothman" its original name (Nickell 2011, 175–181)! Among the barred owl's folk names are Crazy Owl, Round-Headed Owl, Swamp Owl, and many others (Rogers 2008, 84).

Obviously young Davis did not see in the dark that the creature could fly, perceiving only a blur of its wings. He would also see it with its wings down, and—as with the Dover Demon (Nickell 2023)—mistakenly interpret the wings' front edges as long, spindly arms. At the ends of such (shown in Davis's sketch) were its "three big fingers"—actually splayed feathers common to birds at their wingtips. (The portion of a wing that corresponds to the human hand is termed the manus [Latin for hand]. See Alsop 2001, 731; Rogers 2008, 32.) This same effect of "arms" and "fingers" has been reported in the cases of other owl monsters: both the Dover Demon (Nickell 2023) and the Kelly, Kentucky, creatures of 1955 (Nickell 2011, 171). (It underscores my longtime urging that investigators study illusions by becoming magicians.)

Indeed, as to the "blur of green running," this seems to have occurred early, therefore witnessed before Davis got back in the car, or he saw it out a side window. In other words, it was possible from the flapping of wings as the barred owl began flight, disturbing green vegetation. Most interestingly, barred owls are "Territorial all year round, and chase away intruders." In fact, humans may provoke one to "flee, perform a noisy distraction display with quivering wings, or even attack" ("Barred Owls" 2023).[3] We may well suspect that such an attack was what Davis experienced!

In any event, it appears that the sight was a perception in a told-and-retold narrative. The local sheriff, Liston Truesdale, was impressed that Davis intended to relate a true story, and—for what it is worth—he passed a polygraph test without trouble (Radford 2023, 59; Blackburn 2013, 33).

Denouement

As with so many hapless persons who encounter "monsters," the percipient in this case was caught by surprise, frightened, challenged to see something that was difficult to see, confused later as to the sequence of details, grilled by debunkers, half-listened to by sarcastic reporters, and exploited by others hoping for profit.

In my long, if unusual, career, I have come across "monster" witnesses accused of being hoaxers—just like Chris Davis. He is gone now, murdered June 17, 2009, at age thirty-nine, over a marijuana deal gone bad. In the book Lizard Man: The True Story of the Bishopville Monster, Lyle Blackburn (2013) has told us more, perhaps, than we wanted to know about Davis and his encounter. There are accounts by many others, plus earlier sightings, hoaxes, speculations about "underground reptoids," shapeshifting "reptilians" (some with long tails), a frog-like cryptid, "gatorman," the Skunk Ape and other versions of Bigfoot (or what I have renamed "the Bigfoot Bear"—one seen in its upright mode), and so on, Lizard Man festivals, souvenirs, and much, much more, including the "Curse of Lizard Man," and "Lizardmania"—ad nauseum (45, 89–123, 143, 156–157).

Just about the only thing missing is a barred owl! Indeed, Blackburn mentions other swamps where (unknown to him, apparently) barred owls exist. One is the Honey Island Swamp in Louisiana (Blackburn 2013, 107, 154). At that wilderness area I searched for its fabled (originally hoaxed) swamp monster in 2000. My Cajun guide, Robbie Charbonnet, had never seen a monster there, but from merely glimpsing a silhouette, he identified for me a barred owl, then carefully steered his boat through cypresses hung with Spanish moss so I could have a closer view (Nickell 2011, 214–215). I don't know if anyone ever came in the night and saw its "glowing" red eyes, but I once encountered one in the wild that I can see still. Unlike Chris Davis, I knew what it was.

Let us review the elements in Davis's experience that identify the monster he saw as a barred owl: The creature was nocturnal; it inhabited a swampy woodland and exhibited crimson eyeshine; it had scales or seemed caked with mud; it appeared as "a blur of green" (likely a "quivering wings" display in moonlit greenery, presaging an attack); it seemed to have "arms" (front wing edges) with long "fingers" (splayed-feather wingtips), as well as rough-skinned legs, each possessing three toes with "nails" (an excellent description of a barred owl's tarsus—lower, featherless part of the legs—and talons); and the creature was territorial, making an attack on a human intruder. Hopefully, someday if you encounter such a creature yourself in the wild, you won't believe it's a Lizard Man! You'll recognize a barred owl.

Notes

1. The sighting was not "late at night" on June 28, 1988, apparently, as often reported, but "in the early morning hours of June 29" at "around 2:30 a.m.," according to Blackburn (2013, 9, 165).

2. Here, the reference is probably not to the large "fingers" (at wingtips) but instead to the owl's very powerful talons. One writer tells how a barred owl once put a claw through his hand (Rogers 2008, 162).

3. Could a "quivering wings" display in the moonlit greenery have been the source of the "blur of green"?

References

Alsop, Fred J. III. 2001. Smithsonian Handbooks: Birds of North America, Eastern Region. New York, NY: DK Publishing.

Barred Owls. Online at allaboutbirds.org/; accessed April 20, 2023.

Bingham, Joan, and Dolores Riccio. 1991. More Haunted Houses. New York, NY: Pocket Books.

Blackburn, Lyle. 2013. Lizard Man: The True Story of the Bishopville Monster. San Antonio, TX: Anomalist Books.

Bull, John, and John Ferrand Jr. 1994. National Audubon Society Field Guide to North American Birds: Eastern Region. New York, NY: Alfred A. Knopf.

Manley, Roger. 2007. Weird Carolinas. New York, NY: Union Square and Co.

Nickell, Joe. 2011. Tracking the Man-Beasts: Sasquatch, Vampires, Zombies, and More. Amherst, NY: Prometheus Books.

———. 2023. Identifying the enigmatic "Dover Demon." Skeptical Inquirer 47(4) (July/August): 16–19.

Radford, Benjamin. 2023. The Curious Case of the Carolina Lizard Man. Skeptical Inquirer 47(3) (May/June): 58–61.

Rogers, Denny. 2008. The Illustrated Owl: Barn, Barred & Great Horned. Mount Joy, PA: Chapel Publishing Co.

Afterword

We were honored when Joe Nickell approached us to publish this volume as the inaugural release from Monster House, LLC. Monster House is the production company behind podcasts such as MonsterTalk and *In Research Of...*. You can find our audio content on your podcast platform of choice, but we are expanding into print and audiobooks and are delighted to present the first of what we hope to be many works that use monsters to explore the natural world and its many mysteries. Thank you so much for joining us on this journey.

Blake Smith
President, Monster House LLC
Kennesaw, GA
Summer, 2023

Index

A

Abominable Snowman .... *See* Bigfoot

Almas .... *See* Bigfoot

Animals

    Ape 112, 175

    Baboon 163, 166

    Barn Owl 119, 194

    Barred Owl 119, 194, 195, 252–54

    Bear, Black 101, 122–24, 122, 131–39

    Bear, Brown 132, 153

    Catfish 77–85

    Coelacanth 180

    Eel 54, 55

    Gorilla 173

    Manatee 73–74

    Monkey 101, 127, 182, 241

    Narwhal 5, 33–36

    Otter 23, 42, 47, 55, 63, 65, 66

    Panda 180–87

    Panther 175

    Seal 23–27, 41–42, 72–73

    Seal, Elephant 72–73

    Shark 39, 40

    Whale 16, 40, 41

Ape man/men 100, 101

B

Barred Owl 252

Bigfoot

    Abominable Snowman 109

    Alma 99

    Bardin Booger... 148

    Big Muddy 136

    Bigfoot Bear...132, 152, 253

    Bigsuit...................... 91, 110–13, 119, 124, 148, 167, 186

    Fouke Monster... 134

    Momo 134, 136

    Primary Chapters... 93–170

    Sasquatch.............. 91, 94, 100, 102, 105, 107–8, 109, 112

    Skunk Ape 131, 138, 143–51, 253

    Yeren 99, 100–102, 127

    Yeti 99–100, 102

    Yowie 161–68, 174

Bogeyman .... *See* Supernatural Monsters

Buk'wus ....*See* Folklore Creatures

C

Cadborosaurus....................................*See* Water Monsters

Coleman, Loren 161, 194, 233, 243

Cressie..*See* Water Monsters

D

Dermal Ridges 110, 124

D'Sonoqua ....*See* Folklore Creatures

E

Eel 48, 55, 56, 119

F

Feltham, Steve 49

Fingerprints 110, 124

Folklore Creatures

Buk'wus 106, 107

D'Sonoqua 106

Jörmungandr 10

Kelpies 45–46, 48

Kraken 10–16

Kwakiutls 106

Lorelei 206–11

Mermaid 206, 208

Popobawa 212–15

Vrykolakas 218

Footprints 99, 100, 101, 108, 109, 119, 121, 124, 167

Fouke Monster .... See Bigfoot

G

Giant Catfish .... See Water Monsters

Giant Eel .... See Water Monsters

Giant Squid 39, 52

Gimlin, Bob 110

H

Hoax 44, 49, 63, 106, 109, 112, 113, 138, 167

Huevelmans, Bernard 30, 99, 182

J

Jörmungandr ....*See* Folklore Creatures

K

Kelpie...*See* Folklore Creatures
Kraken...*See* Water Monsters, *See* Folklore Creatures
Kwakiutls...*See* Folklore Creatures

L

Lorelei..*See* Folklore Creatures

M

Mackal, Roy 23, 70, 71
Mermaid..*See* Folklore Creatures
Momo ....*See* Bigfoot
MonsterQuest.. 11, 49, 53–57, 231, 234

N

Nessie...*See* Water Monsters

O

Ogopogo ....*See* Water Monsters

P

Paranatural Naturalist 21, 119, 231
Patterson, Roger 91, 110, 119, 167, 186
Popobawa ....*See* Folklore Creatures

S

Sasquatch ....*See* Bigfoot
Sea Ape ....*See* Water Monsters
Sea Serpent ....*See* Water Monsters

Shine, Adrian 46, 48

Skunk Ape .... *See* Bigfoot

Strange Creatures

    Chupacabra 200, 222, 231–37

    Dover Demon 240–47

    Honey Island Swamp Monster 88–91

    Lizardman, Bishopville 250–55

    Montauk Monster.. 198–203

    Mothman 119, 191–95, 252

Supernatural Monsters

    Bogeyman 101, 107, 114, 153, 154, 165

    Mercy Brown 226–29

    Moosham Castle 217–23

    Vampire 226–29

    Werewolf... 217–23

    Werewolf, Becoming One 217–18

V

Vampire .... *See* Supernatural Monsters

Vrykolakas.. *See* Folklore Creatures

W

Water Monsters

    Cadborosaurus 39–42

    Cressie 52–58

    Giant Catfish 77–84

    Giant Eel 11, 53, 54–55

    Kelpie.. 45–46

    Kraken 10–16

Nessie 44–49, 64, 65

Ogopogo 41, 64

Sea Ape 21–27

Sea Serpent 30–36, 39–40, 42, 46, 63

White River Monster 70–75

Werewolf .... *See* Supernatural Monsters

White River Monster .... *See* Water Monsters

Y

Yeren .... *See* Bigfoot

Yeti .... *See* Bigfoot

Yowie .... *See* Bigfoot

www.ingramcontent.com/pod-product-compliance
Lightning Source LLC
Chambersburg PA
CBHW070801280326
41934CB00012B/3003